SpringerBriefs in Latin American Studies

SpringerBriefs in Latin American Studies promotes quality scientific research focusing on Latin American countries. The series accepts disciplinary and interdisciplinary titles related to geographical, environmental, cultural, economic, political and urban research dedicated to Latin America. The Series will publish compact volumes (50 to 125 pages) refereed by a region or country expert specialized in Latin American studies. We offer fast publication time of 8 to 12 weeks after acceptance. The series aims to raise the profile of Latin American studies, showcasing important works developed focusing on the region. It is aimed at researchers, students, and everyone interested in Latin American topics.

More information about this series at http://www.springer.com/series/14332

G. Thomas LaVanchy · Matthew J. Taylor ·
Nikolai A. Alvarado · Anna G. Sveinsdóttir ·
Mariel Aguilar-Støen

Tourism in Post-revolutionary Nicaragua

Struggles over Land, Water, and Fish

CLAG
clagscholar.org

Springer

G. Thomas LaVanchy
Department of Geography
Oklahoma State University
Stillwater, OK, USA

Nikolai A. Alvarado
Department of Geography & Geographic Information Science
University of Illinois Urbana-Champaign
Urbana, IL, USA

Mariel Aguilar-Støen
Centre for Development and the Environment
University of Oslo
Oslo, Norway

Matthew J. Taylor
Department of Geography & the Environment
University of Denver
Denver, CO, USA

Anna G. Sveinsdóttir
Department of Natural Resources & Environmental Sciences
University of Illinois Urbana-Champaign
Urbana, IL, USA

ISSN 2366-763X ISSN 2366-7648 (electronic)
SpringerBriefs in Latin American Studies
ISBN 978-3-030-55631-0 ISBN 978-3-030-55632-7 (eBook)
https://doi.org/10.1007/978-3-030-55632-7

This Springer imprint is published by the registered company Springer Nature Switzerland AG
The registered company address is: Gewerbestrasse 11, 6330 Cham, Switzerland

Acknowledgments

These words exist because the people of Gigante allow them to exist. The people of Gigante, new and old, have allowed us into their lives over the last decade and a half. They have let us ask questions about how they live, what they think about the rapid pace of change due to tourism, where they get their water from, what was their land, what is their land now, how they fought for the revolution, how they came to Gigante, and how they and their children see the community's future. We work for you and with you. Not one word printed here would be possible without Maria, Rufino, and Maximo Lanza Zapata of Gigante. We acknowledge the rest of the population of Gigante with the same gratitude. It is a gratitude that goes far beyond work relationships because we are all now "*compañeros*" fighting for what we see is right, perhaps. As academics, it is our obligation to bring light the struggles of people and the changes taking place to their lives and their resources. *Seguimos, entonces...*

The bringing together of academic and local lives is enabled by money. Yes, we need funds to work with communities to bring their struggles to light! Those funds came from the University of Denver's Public Good fund and PROF fund, the Lawrence Herold fund of the University of Denver's Department of Geography & the Environment, the Conference of Latin Americanist Geographers (CLAG), the Latin American Specialty Group and Development Geographies Specialty Group of the American Association of Geographers (AAG), the Geological Society of America (GSA), and, somewhat surprisingly, the US Environmental Protection Agency (EPA).

Contents

1 Tourism Along Nicaragua's Pacific Coast: Context and Dilemmas ... 1
1.1 The Global Context for Tourism in Playa Gigante ... 3
1.2 Nicaragua's Political and Economic Context ... 4
1.3 Tourism in Nicaragua ... 4
1.3.1 Tourism Development in Gigante ... 6
1.4 Approach ... 8
1.5 Overview ... 9
References ... 12

2 Who Owns the Sea? Conflicts Between Artisanal Fisheries and Tourism ... 15
2.1 Introduction: Tourism, Traditional Livelihoods, and Conservation Schemes ... 15
2.1.1 Political Ecology and Resource Use Struggles ... 17
2.1.2 Conservation Schemes and Resource Control: Marine Protected Areas ... 18
2.2 Results ... 19
2.2.1 Artisanal Fisheries in Gigante ... 20
2.2.2 The Spaces of Fishing in Gigante ... 21
2.2.3 Economics of Fishing in Gigante ... 24
2.2.4 The State of Fish Stocks and the Discourse of Unsustainable Fishing ... 26
2.2.5 Tourism, Local Jobs, and Changes in Artisanal Fishing ... 27
2.2.6 La Anciana Marine Protected Area and Conservation Discourse: Tourism Versus Local Livelihoods ... 28
2.3 Discussion and Conclusions ... 30
References ... 32

3 Tourism as Tragedy? Common Pool Problems with Groundwater 35
3.1 Introduction 35
3.2 Water and Tourism 36
3.2.1 Water in Southwest Nicaragua 37
3.2.2 Geology 38
3.2.3 Water Law and Common Pool Issues 38
3.3 Evidence of Conflict 40
3.4 Discussion 46
3.4.1 Whither Tourism? 47
3.4.2 Paths Forward 48
3.5 Conclusions and Recommendations 49
References 50

4 Imagining the "Emerald Coast": Insecure Land Tenure, Property Disputes, and Tourism Development 53
4.1 Introduction 53
4.2 The Historical Land Context in the Area 54
4.3 Land Conflicts and Tourism Development 55
4.3.1 Playa Amarillo Conflict 56
4.3.2 Caso Tola Conflict 57
4.4 Imagining the Emerald Coast 60
4.4.1 Legislative and Institutional Factors 62
4.4.2 Elite Dynamics 63
4.5 Conclusion 65
References 68

5 From Boom to Bust and Back Again? Current Dilemmas Related to Tourism Along Nicaragua's Pacific Coast 71
5.1 Introduction 71
5.2 Cycles in Tourism 72
5.2.1 Tourism in Its Current State in Gigante 73
5.2.2 Moving On? 74
5.3 Concluding Thoughts 76
References 78

List of Figures

Fig. 1.1 Map of the study area in Nicaragua. Cartography: Mary Lee Eggart 2
Fig. 1.2 Growth of tourism in Nicaragua (1994–2018). *Source* UNWTO (2019) and calculated from data. **a** Millions USD. **b** 1000 5
Fig. 1.3 Getting the story through a local perspective. *Source* G. T. LaVanchy 11
Fig. 1.4 Getting the story with local assistance—collecting water-level data. *Source* M. J. Taylor 11
Fig. 2.1 Map of study area and site for La Anciana marine protected area. Cartography: Mary Lee Eggart 16
Fig. 2.2 The fiberglass boats (pangas) of Gigante's artisanal fisheries are a dominant sight along the beach, showing the importance of this economic activity to the community. *Source* N. A. Alvarado 19
Fig. 2.3 A local fisherman weighs his snapper catch at one of the acopios. Monopolization of gasoline and direct commercial ties with buyers forces fishermen to sell their catch to acopios, giving the middlemen a privileged position in the political economy of Gigante's fisheries. *Source* N. A. Alvarado 22
Fig. 2.4 Access to the shoreline is vital for many fishermen who do not own pangas or who seek to reduce operation costs and debt with acopios. *Source* N. A. Alvarado 23

Fig. 2.5 The different fishing methods used by Gigante's fishermen. **a** The trasmallo used to target lobster rests on the ocean bottom and is placed very close to shore in areas with rocky substrate. **b** The net used to target snapper rests just above the ocean bottom and is placed in areas 10–30 m deep. **c** The snapper longline rests on the bottom of the ocean and is placed in areas known to have underwater rocky formations. **d** The longline to target Dorado floats near the ocean surface (10 m max) and is placed in open waters beginning at around 10 km from the coast. *Source* Alvarado and Taylor 25

Fig. 2.6 Young locals hitch a ride on top of an oxcart carrying trasmallos. These types of scenes portraying local culture could come to an end with increasing tourism and conservation discourses that displace fishermen without carrying serious historical, economic, and environmental studies. *Source* N. A. Alvarado 29

Fig. 3.1 Climograph for Tola region. *Source* G. T. LaVanchy and Instituto Nicaragüense de Estudios Territoriales (INETER) ... 37

Fig. 3.2 Typical hand-dug (artisanal) well in Gigante. *Source* G. T. LaVanchy 39

Fig. 3.3 Proliferation of wells in Gigante. *Source* G. T. LaVanchy 41

Fig. 3.4 Annual rainfall for study area. Dashed horizontal line represents the mean of 1440 mm for the climatic series of 1968–2019. *Source* G. T. LaVanchy, M. J. Taylor, and Instituto Nicaragüense de Estudios Territoriales (INETER) ... 42

Fig. 3.5 Well inventory from June 2015 revealing the relative quantity of water in Gigante wells. Minimum indicates less than 0.5 m remaining; Low indicates less than 1.0 m remaining. *Source* G. T. LaVanchy and Digital Globe imagery 43

Fig. 3.6 Daily hydraulic head values for two drilled well in Gigante showing an ~4 m drop in the water table from 2013 to 2015. *Source* G. T. LaVanchy and M. J. Taylor 43

Fig. 3.7 Transporting water from a distant well to home. *Source* M. J. Taylor 44

Fig. 3.8 Private artisanal wells contaminated from saltwater intrusion. *Source* G. T. LaVanchy and Google Earth 45

Fig. 4.1 The group of ex-combatants who occupy part of the Playa Amarillo area. Here they are seen celebrating the 30th anniversary of "El Triunfo" of the revolution on July 19, 2009. *Source* M. J. Taylor 57

Fig. 4.2 El Bigote of Playa Gigante, Nicaragua. *Source* Madeline Meredith 58

Fig. 4.3 **a** Plans of the Pedro Joaquín Chamorro agricultural cooperative showing individual plots surveyed as part of the land grabbing scheme by the investment group González-Bolaños. **b** Its title, "cessation of the community" can be seen in the zoom view of the plans. This cessation of the community was an illegal move by the land developers and their lawyers, done without permission of the cooperative. *Source* M. J. Taylor 59
Fig. 4.4 Branding and advertising of the Emerald Coast by tourism developers. *Source* G. T. LaVanchy 61
Fig. 4.5 Juan Ramon from Gigante sings "Mi Papa Tenía un Terreno" while Rufino Lanza Zapata (left) reflects and listens. June 2012. *Source* M. J. Taylor 66
Fig. 5.1 Rufino Lanza Zapata. In times of crisis, younger residents have turned to Rufino for advice on cultivating subsistence crops. Rufino is one of the original members of the agricultural cooperative that was granted the land around Gigante in the 1990s. *Source* M. J. Taylor 74
Fig. 5.2 Manca Ruiz has fished the seas around Gigante since the early 1990s. He has been somewhat immune to changes brought about by the decline in tourism because he has always relied on the sea rather than tourism. *Source* Jamey Thomas 76
Fig. 5.3 Fishermen like Manca have incorporated new members into their fishing teams. *Source* Jamey Thomas 77

Map

Map of Nicaragua showing study area

Picture

Playa Gigante, Nicaragua: A Fishing Village no more? *Source* M. J. Taylor

The people of Gigante, like Maximo Lanza Zapata, have allowed us into their lives over the last decade and a half. *Source* G. T. LaVanchy

Chapter 1
Tourism Along Nicaragua's Pacific Coast: Context and Dilemmas

> *and why must we develop tourism in this God-forsaken country? Because it's the path to salvation for the national economy.*
> —Comandante Tomás Borge, 2003 (As cited in Babb 2004, p. 541).

Abstract This chapter provides an overview of tourism as it exists globally and details on the particulars of its emergence and presence in Nicaragua. It also provides a history of the community of Playa Gigante and a description of its natural resources. It concludes with an overview of the contents of this book.

Keywords Tourism · Nicaragua · Pacific coast

Tourism is a defining theme throughout this book, but is not the only point. This book is equally about a place (Playa Gigante) experiencing tourism. How it became a place of tourism interest is both nuanced and straightforward. These points will come to light over the course of each chapter. A helpful point of departure for understanding the influence of tourism in Playa Gigante, and indeed in Nicaragua, is to present the context of global tourism and the narrative of its efficacy as a development tool for progress. Following a contextualization of global tourism, this chapter will present tourism in Nicaragua as a whole, then its emergence along the Pacific coast. Playa Gigante (hereafter referred to as Gigante) lies on the southwest coast of Nicaragua overlooking the Pacific Ocean (Fig. 1.1). There one can find forested hills sloping to various rocky and sandy beaches. This landscape attracts visitors who enjoy the flora and fauna of dry tropical forests, those who appreciate secluded beaches, and especially those looking for waves to surf. But what happens after these visitors arrive? How do they impact the environment? What influence do they have, economically and culturally, on local populations? These questions form the topic of this book.

Although Gigante is the geographical focus of this book, tourism has profoundly shaped other communities along the southwest coast of Nicaragua. Many of these communities share similarities with Gigante with respect to conflicts over land, water, and fishing economies, though some face additional or different challenges

G. T. LaVanchy et al., *Tourism in Post-revolutionary Nicaragua*,
SpringerBriefs in Latin American Studies,
https://doi.org/10.1007/978-3-030-55632-7_1

Fig. 1.1 Map of the study area in Nicaragua. Cartography: Mary Lee Eggart

not covered in this book. Gigante was selected as the focus of research for several reasons, not the least of which was access granted to us as researchers by community members and the openness of the community to share their stories. The research presented in this volume was not conceived in an office far from Nicaragua, but developed in collaboration with community members about issues that they found important in their lives. The authors have maintained reciprocity by frequently conducting community-wide meetings to discuss findings, validate perspectives, and solicit input. In the sharing of this research, the authors recognize that there may be other angles or perspectives to explore, such as the experience by tourists and the stories of other communities not influenced by tourism.

1.1 The Global Context for Tourism in Playa Gigante

Tourism is often proposed as a way to diversify local economies and alleviate poverty and vulnerability endured by rural and coastal communities (Chok et al. 2007; Holden et al. 2011; Spenceley and Meyer 2012; Telfer and Sharpley 2016). Its footprint as a US$1.5 trillion global industry (UNWTO 2019) continues to grow as global commodity chains are strengthened and communication networks grow in their capacity to attract visitors to distant locales. This form of economic stimulus is often an attractive means for increasing national revenues and boosting foreign direct investment in many developing countries (Messerli 2011; Oyewole 2009; Stonich 1998). International donor agencies, such as the World Bank, USAID, and the Inter-American Development Bank, have played key roles in extending this form of development throughout Central America. While the importance of tourism in terms of global services trade and income generation is undeniable, whether these benefits remain within the community and improve the conditions locally is debatable (Stonich 2005; Young 1999). Wilson (2008) pointed out that the benefits of tourism are strongly heralded by those with a neoclassical economic approach, while those with neo-Marxist and dependency approaches focus on its costs. Progress (or regress) through tourism is often in the (ideological) eye of the beholder.

Economic benefits alone may not ensure a socially and culturally sustainable development for local communities. As a result, in order to gauge the success of tourism as a development strategy, scholars have suggested taking into consideration other tourism-related changes at the local scale beyond the economic. Stronza and Gordillo (2008) argued, for example, that tourism-related changes that affect a community's social cohesion, cooperation and reciprocity among individuals, small producer empowerment, and traditional culture, could determine the success or failure of tourism as a development strategy for local communities. These aspects of a community, however, are often drastically transformed with the influx of tourists and as local livelihoods make way to an increasingly tourism-dominated economy (Kottak 1992; Mbaiwa 2011).

Indeed, one of the assumptions embedded in the discourse in favor of tourism development is that local communities will benefit from the creation of jobs as local producers move from subsistence livelihoods to paid employment in the tourism sector (Chok et al. 2007; Solís Librado 2009). However, the creation of jobs represents at best the economic facet of community development, while social, cultural, and environmental facets are largely ignored (Stronza and Gordillo 2008). In the excitement for job generation, development experts and supporters forget that life becomes more expensive in general for locals when tourism capital dominates local markets, resulting in many being displaced. Furthermore, due to the seasonal tendency of the tourism industry and its reliance on the global economy, communities that increasingly rely less on subsistence livelihoods can be left ill-prepared to face economic crises. Thus, a better approach to tourism development should attempt to diversify local livelihoods while safeguarding the subsistence activities that have forged a community's socio-economic and cultural traits.

1.2 Nicaragua's Political and Economic Context

The United States has played a pivotal role in Nicaragua's political and economic life. Because of the geographical proximity that made Nicaragua a "U.S. backyard" and inspired by the Manifest Destiny that drove U.S. imperialistic ambitions since the nineteenth century, Nicaragua became subject to continuous U.S. interventions that manipulated political and economic conditions to make colonial and neocolonial forms of exploitation possible (Gobat 2005; Walker and Wade 2011).

The most significant of these interventions in Nicaragua's recent past was the U.S. support of the Somoza dictatorship beginning in 1936, which lasted for 43 years until the triumph of the Sandinista revolution in 1979. The United States supported the rule of three Somoza family members, disregarding the economic, political, and human atrocities conducted in the name of capitalism. The Frente Sandinista de Liberación Nacional (FSLN) emerged on to the scene in the 1950s as an anti-Somoza/anti-U.S. armed guerrilla movement seeking to return the country to democracy after 43 years of despotism. The revolution succeeded in 1979, and 11 years of Sandinista government followed (1979–1990). Concurrently, a series of reforms aimed to bring the masses out of poverty through the redistribution of wealth and land. Prominent among these reforms was the Ley Agraria (Agrarian Reform Law) of 1981, which redistributed land among rural masses organized in agricultural cooperatives (Everingham 2001; Jonakin 1996). The success of the Sandinista revolution was, however, limited due to U.S. sanctions and a U.S.-backed Contra revolution that slowly bled the country of human blood and money from 1980 to 1990.

The Sandinistas lost power in 1990 to a regime inclined toward the more familiar economic and "democratic" models of "U.S-friendly" nations (Brown 2001). The neoliberal governments of Chamorro, Alemán, and Bolaños governed Nicaragua for the 1990–2006 period until the return of Daniel Ortega and the Sandinista party in January of 2007. The Sandinista party continues to maintain power and is elected until January of 2022. These continuous political transitions in Nicaraguan history perpetuate social, political, and economic turmoil, and created instability in many areas of Nicaraguan life that persist today and that have left marginalized populations vulnerable to further exploitation (Cupples 1992; Hunt 2011; Kinzer 1991; Zimmerer 2000).

1.3 Tourism in Nicaragua

The research for this book begins with the premise that resource usage is a fundamental aspect of tourism activity. Resources such as water, land, and energy are fundamental in making and supporting the tourism industry, although more frequently tourism is framed in terms of economic outcomes. Indeed, global tourism has grown into one of the most dynamic industries of the modern economy. The reality of this economic dynamism captured the interest of government leaders in Nicaragua

	1994	2000	2005	2010	2015	2016	2017	2018
Tourism receipts[a]	40	129	206	313	528	642	841	544
Arrivals[b]	238	486	712	1,011	1,386	1,504	1,787	-
Annual growth in receipts (%)		21.5	9.8	8.7	11.0	21.6	31.0	-35.3
Annual growth in arrivals (%)		12.6	7.9	7.3	6.5	8.5	18.8	-

Fig. 1.2 Growth of tourism in Nicaragua (1994–2018). *Source* UNWTO (2019) and calculated from data. **a** Millions USD. **b** 1000

despite their overall aversion to capitalism as a form of progress. In fact, former revolutionary leaders such as Comandante Tomás Borge saw it as a "path to salvation for the national economy" (Babb 2004, p. 541). In 2018, total international tourism receipts reached US$1,451 billion (UNWTO 2019) and until early 2018, tourism injected the much-needed foreign capital into Nicaragua's economy (Babb 2004; Hunt 2011; Zapata et al. 2011). After years of political and revolutionary turmoil, Nicaragua reinvented itself in the late 1990s as a place with "unspoiled" natural beauty, *gringo*-friendly locals, and ample opportunities to buy, rent, or borrow a slice of paradise (Dear 2013; LaTourrette 2006; Nichols 2010; LaVanchy and Taylor 2015). Babb (2004) and others have detailed the remaking of Nicaragua into a safe and desirable tourist destination, thanks in part to abundant natural resources, neoliberal economic reforms, and endorsement from satisfied travelers. Indeed, since the mid-1990s, tourism arrivals and receipts have grown at a steady rate (Fig. 1.2).

Prior to the beginning of the tourism boom in the early 2000s, arrivals to Nicaragua were tourists from other Central American countries or backpacker, budget-minded travelers from North America and Europe (INTUR 2009). This mode of tourism produced an average of $325 USD per tourist visit, the lowest in Central America and one-third that of Guatemala and Costa Rica (Vargas et al. 2010). This resulted in renewed effort by the government and investors to promote tourism in a variety of forms and durations. In 1999, the Nicaraguan government increased its commitment to and dependence on the foreign-exchange earnings from tourism by passing Law 306, making tourism an "industry of national interest" (Article 1). This was further codified through a reform bill (Law 495) in 2004 that offered aggressive incentives and benefits to individuals or corporations, both national and foreign, which generate tourism activities.

This prioritizing by the national government translated into continued increase in tourist arrivals and diversification in types of tourism that some credit for the concomitant increase in tourism receipts. Perhaps, the most pronounced example of this diversification was the rise in residential tourism. Matteucci et al. (2008) documented the increase in second home tourism proliferating along the southwest coast in the early 2000s. Although these "holiday" homes often lack full-time residents, they represent a significant injection of revenue into local economies. A significant portion of residential tourism in Nicaragua has agglomerated along the southern Pacific coast where population densities are low and weather is more temperate and seasonally dry.

During the late 1990s, the southwest coast of Nicaragua was particularly attractive to those searching for empty waves, pristine coastline, and the pace of unhurried life (Alvarado and Taylor 2014; LaVanchy and Taylor 2015). Initial foreign exploration of this region of Nicaragua was speared by a brand of surfers with minimal concern for amenities and deep desire for adventure on previously undiscovered surf breaks. This opened up a frontier for other types of tourists and investors and led to growth in infrastructure and amenities. Thanks to outside influence, a burgeoning surf culture now exists along the Pacific coast that supports all level of surfers—both foreign and domestic. Nicaragua's tourism authorities (INTUR) have been quick to recognize this boom and have shaped media campaigns to include hosting of major surf competitions such as the 2012 World Masters and 2013 Junior World Championships along Nicaragua's southern Pacific coast.

Tangential to the growth of surf tourism has been an expanding interest in upscale homes and a wider range of amenities (i.e. spas, yoga retreats, restaurants, and adventure sports). Today's surf tourist is just as apt to purchase a second home as they are to shack up in a hostel for two months. Budget-minded tourist still trek to the southwest coast, but well-heeled surfers and urbanites from Managua are finding their own reasons for pilgrimage to the coast. This opening up of the coast has been regarded by many as a positive change because of increased employment opportunities; however, there is no denying that it has come at a cost with respect to resources and culture. The topic of this book thus explores what could happen in Nicaragua through the lens of Gigante. Indeed, further to the south in San Juan del Sur, arguably the first place for beach tourism in Nicaragua, the benefits of tourism development at the community level are debatable. For example, in San Juan del Sur, Egan (2011) showed that foreigner-driven short-term and residential tourism has changed the landscape by making land and property unaffordable to locals and by dismantling traditional systems of livelihood. The increased influx of foreign capital to San Juan del Sur has not translated into a more equitable distribution of wealth and it "threatens to create neocolonial spatial, economic and social dualism" (Egan 2011: 3). Moreover, Babb (2010) documents an increase in prostitution associated with the rise in tourism over the decade of 2000–2010. An additional impetus of this book, then, is to question, *what will become of Gigante as it emerges as a major destination?*

1.3.1 Tourism Development in Gigante

The area known today as Gigante was part of a vast private hacienda known as finca Güiscoyol owned by Nicaragua's former president Anastasio Somoza in 1937. From that time until the triumph of the revolution in 1979 and the resultant confiscation and redistribution of the land, the Somoza family used finca Güiscoyol primarily for cattle ranching. The Sandinista government re-distributed the confiscated finca to rural citizens grouped in agricultural cooperatives. The area of Gigante (840 hectares total) was granted to fishermen and small farmers that came together under Cooperativa Pedro Joaquin Chamorro between 1984 and 1987. In 1994, the cooperative finally received a title for the entire area that was granted to them during Sandinista rule (1979–1990).

Subsequently, the land around Gigante has been sold and bought in cycles as various individuals, entrepreneurs, and developers have discovered its natural beauty and tourism potential. Currently, its nearly 800 inhabitants consist of fisherman, farmers, small-business owners, expatriates, and a growing number of migrant squatters looking to improve their economic condition by laying claim to land still contested from the war (nearly 40 such families have settled in the area since 2013). Small-scale commercial and subsistence fishing are the most important economic activities in the community and local culture reflects the community's close relationship with the ocean over the decades. The municipality of Tola (Gigante is in Tola) is starting to experience tourism development similar to that of its southern neighboring municipality of San Juan del Sur. The scale and pace of tourism development in Tola is unprecedented, particularly with the construction of Guacalito de la Isla, a low-density, high investment project that brings wealthy tourists to the area. Guests were whisked from Managua via helicopter or land private jets at a newly constructed regional airport. In Gigante, tourism has proliferated through a transfer of land from local peasants into the hands of developers. Private developers have secured, and continue to secure, more of this coveted coastal property. The demand for land and the large amount of money involved resulted in both judicial and armed conflicts (Bonilla and Mordt 2008). Because land is central to the tourism boom that took place Gigante, one of the chapters in this book delves deeply into the shifts in ownership and conflicts over land in the Gigante area.

Most of Gigante's coastline has developed through the arrival of tourism in the last decade. Historically, the two dirt roads leading to Gigante's coastline constrained the number of arriving tourists. However, the paving of road from the town of Tola to Gigante in 2016 allowed tourism to develop more rapidly thereafter. Despite historically limited access, a number of foreigners have been able to acquire land rights on the coast to build lodging facilities and run marine-based tourism activities. Locals who already resided on the coast are trying to capture tourism revenues through food services, surf taxis, and lodging.

As the international surf community became aware and interested in the idyllic surfing conditions around Playa Gigante, small surf lodges sprang up to accommodate the interest of well-heeled surfers looking for empty waves and the chance for adventure. Over time, news spread about the opportunities for world-class surfing near Gigante and tourism arrivals and expenditures increased (see Fig. 1.2). With this growth came the development of a hostel, a few small hotels, and several restaurants. Additionally, interest in buying land and building vacation homes led to the establishment of several gated communities, poised exquisitely along prime coastline. These larger gated communities exemplify the residential tourism outlined by Matteucci et al. (2008) and reflect the hard, pioneering work of both foreign and Nicaraguan ingenuity, capital, and vision. For the most part, these larger tourist enterprises market themselves as environmentally conscious and supporters of local socio-economic progress. Undoubtedly this pioneering effort has "opened" up the southwest coast to successions of "development" aimed at a range of tourism opportunities, local job productions, education initiatives, and health brigades. This opening up of the coastal face of the Pacific, however, has not come without a struggle and conflict over land (Abu-Lughod 2000; Ferrando 2007; Morales 2007).

Infrastructure and opportunity have grown in tandem with tourism development in Playa Gigante and have created a "pull" effect, attracting Nicaraguans from other parts of the country. Some of these migrants have been employed by or are owners of tourism enterprises, while others have squatted on land to stake a life in fishing or establishing ownership on land that might someday be worth much more as the coast is further commodified. Integrative evaluation of the impact of tourism has yet to be realized due to the nascent nature of its presence in this area. Some researchers and small non-governmental organizations are beginning to capture this change through social, economic, or environmental lenses, though few in-depth studies exist. This book examines some of the impacts of tourism on the land and livelihoods.

1.4 Approach

Research for the chapters covered in this book occurred over the last ten years and involved various aspects of the physical and human geography of the study area. Collectively, they centered on understanding the relationship between tourism development and local populations in southwest Nicaragua through the lens of fishing, groundwater, and land. Particular attention was paid to the political and economic institutions in which tourism is embedded, both globally and in Nicaragua. An array of methods was used in data collection. These included in-depth and informal interviews with local residents, community leaders, tourists, tourism developers, and local NGOs; well monitoring; geological field mapping; and archival work. Throughout each chapter, the discursive and material dimensions of tourism are mapped on to resources fundamental to the livelihoods of individuals tied physically and ideologically to Nicaragua's revolution. Thus, tourism is shown to provide a mechanism for re-accumulating resources such as water and land into the hands of those with more power or money. Equitable access to resources constituted a major theme during the revolution, thus the ability of tourism to facilitate accumulation of resources into the hands of a few proves problematic for obvious social, political, economic, and historical reasons.

1.5 Overview

This book is organized into five chapters, the first of which is this introduction, overview, and context provided by LaVanchy and Taylor. The following two chapters address common pool resources—the ocean and groundwater. In Chap. 2, Alvarado and Taylor describe the historical connection of Gigante's residents to the ocean, their dependency upon its resources, and conflicts between artisanal fisheries and the growing tourism industry. The authors argue that a proposed Marine Protected Area would undermine the livelihoods of fishing families who have traditionally relied on the fecundity of the ocean. Under the narrative of conservation, these original stewards of the ocean would be excluded by more powerful stakeholders within the tourism industry and government. This post-revolutionary re-accumulation of resources by elites comes at an expense to many who fought in Nicaragua's revolution.

In Chap. 3, LaVanchy and Taylor use a political ecology approach to shed light on the nexus of tourism, politics, and access to groundwater. Although this chapter could be the topic for a whole book, LaVanchy and Taylor do a thorough job of outlining the hydrogeology of Gigante, the role of water in tourism, and how the lack of government oversight creates an unsustainable platform for tourism development and local equity. Given the pervasive nature of global tourism and its concomitant need for water, this chapter will be of interest to anyone wanting to understand more on the nexus of tourism and water.

In Chap. 4, Sveindsdóttir and Aguilar-Støen focus on a core resource emerging from the revolution—land. Here, they detail how land is effectively taken back from revolution beneficiaries through a pretense of tourism development and progress. Anyone interested in the discourse of land grabbing in Latin America and the Caribbean will find this chapter salient.

In Chap. 5, Taylor and LaVanchy provide a current description of tourism in Gigante. Since 2018, the national government has demonstrated a heavy hand toward protesters and foreign media who have tried to make public many of the country's social and environmental injustices resulting from the leadership of the new FSLN. The impact on tourism is undoubtedly dramatic. Where previously tourism benefitted through the promotion and endorsement of the national government, now it languishes because of the unsafe image its current political unrest projects to the outside world. What does this mean for local residents in areas where tourism previously flourished? Simultaneously, they lose access to employment while also, potentially, finding renewed access to groundwater, fisheries, and land. Chap. 5 also provides a brief conclusion to the book and thoughts about the future of tourism in Gigante and along the southwest coast.

Readers of this book with prior knowledge of Nicaragua will likely recognize parallels and connections to other studies on Nicaragua. Most notable is the seminal work of Bernard Nietschmann during the 1960s and 1970s with indigenous coastal people along the eastern coast of Nicaragua. *Between Land and Water* Nietschmann (1973) detailed the activities and ecology of the Miskito subsistence system, while *Caribbean Edge* (Nietschmann 1979a) demonstrated the teleconnections between exogenous neoliberal markets and local marine resources. Despite a number of differences, the similarity of local resources under pressure from outside forces is rather stark. This relationship is perhaps most obvious in his article "Ecological Change, Inflation, and Migration in the Far Western Caribbean," which succinctly linked the social and physical changes in a once-isolated community to larger global "currents" (1979). Nietschmann's work was notably grounded in sustained, field-based research and yielded meaningful relationships, agency to local voices, and illuminative scholarship on geographical convergences. We hope our work is a faithful rendering of Nietschmann's approach and influence. In a nod to the commonality in topic and approach, one of the authors was supported by the Bernard Nietschmann Award, a doctoral research grant from the Conference of Latin Americanist Geographers.

Other significant scholarship serves as a base for or reflects symmetry with the focus of this book. Florence Babb (2010) brought attention to the re-emergence of tourism in Nicaragua after the Sandinista Revolution and its subsequent influence on the communities of Managua, Granada, and San Juan del Sur. Julie Cupples shed light on various aspects of neoliberalism in post-revolutionary Nicaragua, which serve as context for this study on tourism (1992, 2004, 2005, 2011). Elsewhere in Latin America, work by Kottack (1992) on social change in a Brazilian village and by Budd (2008) on conflict over water resources in an agricultural valley in Chile provide relevance and a broader context for the changing social landscape of Nicaragua due to outside influence. Other scholarship is noteworthy and are mentioned in the context of the resources covered in subsequent chapters in this book.

The five authors of this book originate from five different countries and bring a diverse range of perspectives and methodologies to the research undergirding this text. Two of the five are women. Still, we recognize our position of privilege, most obvious the opportunity and resources for advanced studies and mobility. This positionality affords a degree of power, which we have been mindful of as we engaged with community members in Gigante. Our goal has been to hear their collective stories and experiences, and to re-tell them through our academically informed lenses of understanding (Figs. 1.3 and 1.4). We trust we have been faithful to their voices.

Fig. 1.3 Getting the story through a local perspective. *Source* G. T. LaVanchy

Fig. 1.4 Getting the story with local assistance—collecting water-level data. *Source* M. J. Taylor

References

Abu-Lughod D (2000) Failed buyout: land rights for Contra veterans in postwar Nicaragua. Latin Am Perspect 27(3):32–62

Alvarado NA, Taylor MJ (2014) ¿Del mar quién es dueño? Artisanal fisheries, tourism development and the struggles over access to marine resources in Gigante Nicaragua. J Latin Am Geogr 13(3):37–62. https://doi.org/10.1353/lag2014.0041

Babb FE (2004) Recycled *Sandalistas*: from revolution to resorts in the new Nicaragua. Am Anthropol 106(3):541–555

Babb FE (2010) The tourism encounter: fashioning latin American nations and histories. Stanford University Press, Stanford, CA

Bonilla A, Mordt M (2008) Tourism PRISMA, Avance de Investigatión No. 4

Brown TC (2001) The real Contra war: highlander peasant resistance in Nicaragua. University of Oklahoma Press, Norman

Budd J (2008) Whose Scarcity? The Hydrosocial Cycle and the Changing Waterscape of La Ligua River Basin, Chile. In: Goodman MK, Boykoff MT, Evered KT (eds) Contentious geographies: environmental knowledge, meaning, scale. Ashgate, Farnham, Surrey, UK, pp 59–68

Chok S, Macbeth J, Warren C (2007) Tourism as a tool for poverty alleviation: a critical analysis of 'Pro-Poor Tourism' and implications for sustainability. Current Issues in Tourism 10(2–3):144–165. https://doi.org/10.2167/cit303

Cupples J (1992) Ownership and privatisation in post-revolutionary Nicaragua. Bull Latin Am Res 11(3):295–306

Cupples J (2004) Rural development in El Hatillo, Nicaragua: Gender, neoliberalism and environmental risk. Singap J Trop Geogr 25(3):343–357

Cupples J (2005) Love and money in an age of neoliberalism: gender, work, and single motherhood in postrevolutionary Nicaragua. Environ Plann A 37(305):322. https://doi.org/10.1068/a37115

Cupples J (2011) Shifting networks of power in Nicaragua: relational materialisms in the consumption of privatized electricity. Ann Assoc Am Geogr 101(4):939–948. https://doi.org/10.1080/00045608.2011.569654

Dear T (2013, June) Nirvana in Nicaragua. *Colorado Avid Golfer*, 28–36

Egan C (2011). Transnationalism: North-to-South migratory flows and their developmental impact in San Juan del Sur, Nicaragua (Masters thesis, Utrecht University, 2011). Utrecht University Repository

Emily H. Young, Human Ecology 27 (4):581-620

Everingham M (2001) Agricultural property rights and political change in Nicaragua. Latin Am Politics Soc 43(3):61–93

Ferrando JE (2007) Consultoría para la elaboración de un studio catastral en el litoral marino de Tola. Informe Final Abril 2007 (MASRENACE: GTZ)

Gobat M (2005). *Confronting the American dream: Nicaragua under U.S. imperial rule.* Durham, NC: Duke University Press

Holden A, Sonne J, Novelli M (2011) Tourism and poverty reduction: an interpretation by the poor of Elmina Ghana. Tourism Plann Dev 8(3):317–334. https://doi.org/10.1080/21568316.2011.591160

Hunt C (2011) Passport to development? local perceptions of the outcomes of post-socialist tourism policy and growth in Nicaragua. Tour Plan Dev 8(3):265–279. https://doi.org/10.1080/21568316.2011.591155

INTUR (2009) Boletín de Estadísticas de Turismo No. 20. Instituto Nicaragüense de Turismo. Retrieved February 2, 2014 from http://www.intur.gob.ni/DOCS/ESTADISTICAS/Estadisticas%20de%20Turismo%202009.pdf

Jonakin J (1996) The impact of structural adjustment and property rights conflicts on Nicaraguan agrarian reform beneficiaries. World Dev 24(7):1179–1191

Kottak CP (1992) Assault on paradise: social change in a Brazilian Village, 2nd edn. McGraw-Hill Inc., New York

Kinzer S (1991) Blood of brothers: life and war in Nicaragua. Putnam, New York
LaTourrette D (2006) Land grab. Surfer's J 15:106–119
LaVanchy GT, Taylor MJ (2015) Tourism as tragedy? Common problems with water in post-revolutionary Nicaragua. Int J Water Resour Dev 31(4):765–779. https://doi.org/10.1080/07900627.2014.985819
Matteucci X, Lund-Durlacher D, Beyer M (2008) The socio-economic and environmental impacts of second home tourism: The South Pacific Coast of Nicaragua example. In: Keller P, Bieger T (eds) Real estate and destination development in tourism: Successful strategies and instruments. Erich Schmidt Verlag GmbH & Company, Berlin, pp 149–161
Mbaiwa J (2011) Hotel companies, poverty and sustainable tourism in the Okavango Delta, Botswana. World J Entrepreneurship, Manag Sustain Dev 7(1):47–58. https://doi.org/10.1108/20425961201000030
Messerli H (2011) Transformation through tourism: harnessing tourism as a development tool for improved livelihoods. Tour Plan Dev 8(3):335–337
Morales VA (2007, Junio 25) Hay pruebas suficientes que demuestran acciones mafiosas del clan González-Bolaños. *La Gente.* Available at http://www.radiolaprimerisima.com/noticias/15851/hay-pruebas-suficientes-que-demuestran-acciones-mafiosas-del-clan-gonzalez-bolanos. Accessed 5 May 2014
Nietschmann B (1973) Between land and water. Seminar Press, New York
Nietschmann B (1979a) Caribbean edge: the coming of modern times to isolated people and wildlife. The Bobbs-Merrill Company Inc, Indianapolis, IN
Nietschmann B (1979b) Ecological change, inflation, and migration in the far Western Caribbean. Geogr Rev 69(1):1–24
Nichols N (2010, March) Nicaragua, the New Frontier. *DMagazine.* Retrieved November 10, 2013, from http://www.dmagazine.com/Home/D_Magazine/2010/March/Travel/Nicaragua_the_New_Frontier.aspx
Oyewole P (2009) Prospects for Latin America and caribbean region in the global market for international tourism: a projection to the year 2020. J Travel Tourism Market 26:42–59
Solís Librado C (2009) Biodiversity and tourism as development alternatives for indigenous peoples. In: Deere CD, Royce FS (eds), Rural social movements in Latin America: organizing for sustainable livelihoods. Oxford: Oxford University Press. https://doi.org/10.5744/florida/9780813033327.003.0012
Spenceley A, Meyer D (2012) Tourism and poverty reduction: theory and practice in less economically developed countries. J Sustain Tourism 20(3):297–317. https://doi.org/10.1080/09669582.2012.668909
Stonich S (1998) Political ecology of tourism. Ann Tourism Res 25(1):25–54
Stonich S (2005) Enhancing community-based tourism development and conservation in the Western Caribbean. NAPA Bull 23:77–86
Stronza A, Gordillo J (2008) Community views of ecotourism. Ann Tourism Res 35(2):448–468. https://doi.org/10.1016/j.annals.2008.01.002
Telfer DJ, Sharpley R (2016) Tourism and development in the developing world, 2nd edn. Routledge, New York, NY
UNWTO (2019) United Nations World Tourism Organization Tourism Highlights, 2019th edn. World Tourism Organization, Madrid
Vargas IV, Navarro CD, Guillén NO (2010) Estudio de Imagen y Posicionamiento del destino de Centroamérica en Europa. In Caballero JLJ, Ruiz PF (eds) *Nuevas perspectivas del turismo para la próxima década,* pp. 601–620 (Sevilla, Espana: III Jornadas de Investigatión en Turismo)
Walker TW, Wade CJ (2011) Nicaragua: living in the shadow of the eagle. Westview Press, Boulder, CO
Wilson TD (2008) Introduction: The impacts of tourism in Latin America. Latin American Perspectives 35:3–20

Zapata MJ, Hall CM, Lindo P, Vanderschaeghe M (2011) Can community-based tourism contribute to development and poverty alleviation? Lessons from Nicaragua. Current Issues Tourism 14(8):725–749. https://doi.org/10.1080/13683500.2011.559200

Zimmerer M. (2000) Sandinista: Carlos Fonseca and the Nicaraguan Revolution. Duke University Press, Durham, NC

Chapter 2
Who Owns the Sea? Conflicts Between Artisanal Fisheries and Tourism

Abstract Since the mid-1990s Nicaragua has turned to the undeveloped beaches of the Pacific to generate much-needed foreign exchange. This region, though, was not uninhabited. Indeed, hundreds of fishing families relied on the fecundity of the ocean to feed their families and to sell fish for international consumption. The growth of tourism brings threats to this form of living. Specifically, marine protected area (MPA), promulgated by large tourism operations, may exclude fishermen from their most fertile fishing waters. This chapter provides details about current use of the ocean by local fishermen and how that will be impacted by the implementation of the MPA.

Keywords Nicaragua · Tourism · Marine protected areas · Artisanal fisheries

2.1 Introduction: Tourism, Traditional Livelihoods, and Conservation Schemes

"Those fishing lines could damage our *trasmallo* (gill net) if we drop it here. Let's move to a different spot by those rocks over there", said Adolfo to the rest of the crew as they scouted the area near La Anciana Island for a good place to set the last net of the day. A high-speed boat carrying foreign tourists and trolling around the island became an impediment to placing the *trasmallo* in this productive fishing spot. Adolfo feared that the trolling lines from the sport-fishing boat would become entangled with and ruin their *trasmallo*—one of their main instruments of subsistence. Meanwhile, beers in hand, the tourists waved and took pictures as if the barefooted fishermen in their modest panga (small fiberglass boat) were part of the excursion they had booked. While these tourists meant no harm, they were oblivious of the fact that their leisure activity had prevented the fishermen access to the resource that they depend on.

Adapted with permission of the Conference of Latin Americanist Geographers by N.A. Alvarado from the article "¿Del mar quién es dueño?: Artisanal Fisheries, Tourism Development and the Struggles over Access to Marine Resources in Gigante, Nicaragua," published in Journal of Latin American Geography, 13(3), 37–62 (2014).

G. T. LaVanchy et al., *Tourism in Post-revolutionary Nicaragua*,
SpringerBriefs in Latin American Studies,
https://doi.org/10.1007/978-3-030-55632-7_2

Fig. 2.1 Map of study area and site for La Anciana marine protected area. Cartography: Mary Lee Eggart

For generations, the seas around La Anciana Island on Nicaragua's southwestern Pacific coast have been one of the most important fishing grounds for the fishermen of Gigante. Now, their access to this area is being threatened by tourism development along the coast and the creation of a marine protected area (MPA) that will enclose and prevent access to these crucial seas and resources (Fig. 2.1). A view of these emerging conflicts comes from Rafael, a local fisherman.[1] He states simply and emphatically, "Del mar quién es dueño? Dígame usted, del mar quién es dueño?" (Who owns the sea? Tell me, who owns the sea?; Rafael 2012). He repeated the statement several times as if it conveyed the only rational answer to the dilemma: to Rafael the ocean does not have property titles like land does; to Rafael the ocean provides access to resources that has supported the livelihood of poor coastal communities for generations. Certainly, Rafael argued, the Sandinista party that he had fought for during the revolution and Contra war would not permit a violation of the very same principles that so many fought for. That is, in his eyes the Sandinistas would not allow a takeover of resources by the new elite.

Taking Rafael's question about ownership of the sea commons as a starting point, this chapter focuses on current and potential impacts of tourism development to

[1]All names of interviewees have been altered to protect confidentiality.

Gigante's artisanal fisheries.[2] In addition, the chapter explores future struggles over access to marine resources that may result from the creation of La Anciana marine protected area, a conservation area being pushed forth by stakeholders in the tourism industry and that spatially overlaps with some of the most productive fishing grounds used by Gigante's fishermen. Findings show how Adolfo's dilemma with the sport-fishing boat is more than a fortuitous decision made in the face of an unusual encounter. His dilemma reflects the future of Gigante's artisanal fishing tradition as it barrels toward collision with the competing interests of capitalism's latest form of intervention in Nicaragua—tourism.

2.1.1 *Political Ecology and Resource Use Struggles*

Patterns of resource use and control are largely shaped by past and current economic models, the political structure put in place to ensure their hegemony over other alternatives, and the social relations forged by these interactions (Blaikie and Brookfield 1987; Bryant 1992; Neumann 2005; Peet and Watts 2004; Robbins 2012). Consequently, existing livelihoods, as well as emerging struggles over resource access, use, and control have been shaped and continue to be shaped by the economic, political, and social context in which they exist. Studies using a political ecology framework are not limited to terrestrial resources and, in the last two decades, several scholars have focused on coastal communities that depend on small-scale fishing for their subsistence (Cinti et al. 2010; Derman and Ferguson 1995; Pomeroy et al. 2007; St. Martin 2001; Young 1999). Coastal fishing communities hold some of the most vulnerable groups of producers in the world, persistently living in poverty and relying on subsistence and small-scale commercial fishing for food and income. Artisanal fishermen also face the threat of declining global fish stocks as a result of decades of predatory industrial fishing (FAO 2012; Mansfield 2011; Béné 2003).

Efforts to mitigate the socio-economic and environmental impacts of predatory large-scale fishing practices have often relied on management schemes that target small-scale fishing efforts and are orchestrated by numerous actors such as international development agencies, NGOs and national governments (FAO 2007). Small-scale, artisanal fisheries around the world have seen their fishing practices and access to spaces regulated as part of this "expert" solution to declining fish stocks and environmental degradation. The expert discourse that supports these schemes considers fishing to be the main driver for fish stock status worldwide and portrays communal holding of fishing space as a recipe for total resource depletion—a tragedy that can only be prevented through private management of the resource base (Hall et al. 2015). Fishing quota is exemplar in this type of management scheme (Mansfield 2007). These types of policies not only ignore success stories of

[2]By artisanal fishing we are referring to any fishing effort lacking mechanized equipment and generally conducted in small boats close to shore. It includes small-scale commercial fishing and subsistence fishing, both intended to support the household economy.

communal management of the commons but also imply that fishermen live in a political, social, and economic vacuum. Consequently, management schemes limit resource use and control by local producers but leave unchanged and unchallenged the socio-economic and political apparatus from which current resource use patterns stemmed in the first place (Hilborn et al. 2004). In addition, management schemes ignore other causes of environmental degradation beyond the influence of local producers such as fish stock declines through industrial fisheries (Mansfield 2011) and stock migration due to changes in the ecosystem (Badjeck et al. 2010; St. Martin 2001). Often, the end result is a disempowerment of local fishermen, who may not, in the first place, be responsible for environmental degradation, but are the easiest to blame. Ultimately, the top-down management strategies tend to transfer resource control to powerful stakeholders that privilege non-extractive (and often very profitable) resource use to the detriment of local fishermen (Mansfield 2011).

2.1.2 Conservation Schemes and Resource Control: Marine Protected Areas

The regulation of livelihood-supporting spaces to promote conservation and sustainability has resulted in numerous conflicts between local producers and other stakeholders. Conservation schemes can be used as power instruments to advance capitalist interests in non-extractive and non-traditional uses (Goldman 2011; Robbins 2012). In addition, meager economic benefits of conservation schemes to local communities may not ensure the success of their implementation. Social and cultural impacts of conservation schemes may also obstruct the internalization of conservation efforts by the communities involved and trigger unsustainable resource use practices (Goldman 2011; Ho et al. 2012; Robbins 2012).

In coastal and marine settings, these conservation schemes take the form of marine protected areas (MPAs). Despite wide use of MPAs as conservation and fisheries management strategies, a conspicuous lack of success stories indicates problems in their effectiveness to meet their proposed goals. For MPAs to meet their proposed ecological goals, for example, studies need to be conducted to ensure that a site is not randomly selected and that the proposed ecological functions of the MPA will indeed take place given the conditions of a particular marine system (Chuenpagdee et al. 2013). In MPAs that are introduced to manage fisheries, the risk of failure is great when the fisheries itself is not well understood (Stevenson et al. 2013). Issues that can cause an MPA to fail in successfully managing fisheries include little understanding of fish mobility and larval dispersal patterns, both of which can result in decrease in yield for fishermen and unsustainable resource exploitation elsewhere from spatial shifts in fishing effort (Hilborn et al. 2004; Stevenson et al. 2013). Lastly, exclusion of local producers and of local knowledge during the initial planning stage can also result in the failure of MPAs to meet their proposed goals (Chuenpagdee et al. 2013; Gonzalez and Jentoft 2011; Kareiva 2006).

Lastly, MPAs can function, like other conservation schemes, to covertly ensure control of resources for non-extractive and highly profitable uses (e.g. for tourism activities; Stonich 2003). As such, MPAs can become neoliberal ocean spaces of commodified resources, ready to be consumed by participants of global capitalistic endeavors. As a resource control strategy, MPAs can directly impact local fishermen by enclosing livelihood-supporting areas (see Christie 2004; Stonich 2003). Who ends up controlling the resources enclosed within the boundaries of MPAs becomes a socio-political issue in which the most powerful stakeholders steer outcomes to their favor during the decision-making process, easily excluding politically weak local fishermen (Chuenpagdee et al. 2013; Stonich 2003).

2.2 Results

In the next few sections we provide findings that highlight the state of artisanal fishing in Gigante and how it is interacting with the tourism boom. First, we illustrate that Gigante is a fishing-dependent community that relies on fishing for food security and income generation (Fig. 2.2). Documenting and understanding how fishermen use marine spaces today, and in the past, permits an understanding of how fishermen are impacted by competition from tourism activities and any restrictions put in place by the MPA. Second, we discuss how fishermen depend on near-to-shore waters for their livelihood. Third, we present the obstacles that prevent fishermen from incorporating tourism-related activities to their livelihood strategy. The obstacles include tourist

Fig. 2.2 The fiberglass boats (pangas) of Gigante's artisanal fisheries are a dominant sight along the beach, showing the importance of this economic activity to the community. *Source* N. A. Alvarado

perceptions of fishermen and the inability to produce surplus from fishing to invest in some sort of transition to tourism. Lastly, we highlight how large tourism operators are using discourses of declining fish stocks and the need for conservation to undermine artisanal fishing and implement a marine protected area that will displace fishermen from some of the most fertile waters along the coast.

2.2.1 Artisanal Fisheries in Gigante

The marine ecosystems of the Pacific southwest coast of Nicaragua act as important nurseries for a number of species of commercial interest, including Snappers (*Lutjanus* spp.), Groupers (*Serranidae*), and the Pacific Green Lobster (*Panulirus gracilis*). The exceptionally high productivity levels along Nicaragua's Pacific coast (see Pennington et al. 2006) provides work for 30% of the country's artisanal fishing force, which accounted for 75% of total fish landings by volume in 2011 (INPESCA 2012). Close to 100% of fishing on the Pacific coast is conducted by artisanal fishermen, illustrating the importance of this activity to the majority of the communities along the coast. In Gigante, artisanal fisheries provide food for the household as a well as a source of income through the harvesting of species of commercial value for international markets.

Gigante's small-scale commercial operation, like most others along the Pacific coast, is mostly intended to supply global demand for fish and lobster. International commercialization of fish catches was expedited by the implementation of The United States-Dominican Republic-Central America Free Trade Agreement in Nicaragua in 2006, putting artisanal fisheries in the loop of the international fish market. According to INPESCA (2012), in 2011, Nicaragua exported 96% of its total fishing production of which 44.27% was exported to the United States, and 36% to Europe. Thus, current resource use patterns among artisanal fishermen in Gigante are largely determined by international demand for valuable species that inhabit the waters near the coast. Besides supplying the global demand for fish, the capture of species of commercial value is more recently also serving to supply increasing demand by local tourism operators. This commercialization to meet distant and local markets is an important income source for many households in the community.

The importance of targeting commercially valuable species to sustain local livelihoods and, consequently, the importance of the coastal waters where these species are caught is illustrated by the prices paid to fishermen for their catch. Although prices fluctuate with global demand, in general in 2012 and 2013, acopios paid fishermen 1.50 USD/lb. for snappers, 1.20 USD/lb. for groupers (Cabrilla), and 1.25 USD/lb. for Dorado (*Coryphaena hippurus*). Acopios pay fishermen 9.50 USD/lb. for lobster tails that weigh 4 oz. and above, and 8.30 USD/lb. for lobster tails weighing between 3 and 4 oz.[3] The income generated by these sales gives fishermen the monetary

[3]The local increase in tourism, however, has created a market for the smaller, restricted lobster tail weighing less than 3 oz, for which acopios pay an average of US$2/lb. Although catching and

capacity to buy alimentary products to supplement their subsistence activities, as well as material products to improve their overall well-being.

Fish also represents an important food source for local households. By-catch from daily operations that target income-generating species is normally taken home for the family to consume. Fishing families rely heavily on fish for food security because many do not own land or farm animals. The importance of fish to local households is evident in the words of Ana, the wife of a fisherman, who stated, "Fish is what we eat. If we can't rely on the ocean to feed ourselves, then, we are going to starve to death" (Ana 2013). Moreover, the network of people that benefit from fisheries is much larger than those involved directly with the fishing labor, however. For example, it is also common for fishermen to give away leftovers such as lobster heads that are used for making soup and other by-catch to friends and family who wait for every boat to return from a day out at sea. In reciprocity for the free food, locals help fishermen pull the pangas out of the water or to carry the outboard motors, gas tanks, and catches to the acopios. By-catch can also be sold to the acopios to generate additional revenue. This low-priced chatarra or "junk" is made of fish species that lack value in international markets. Chatarra is commercialized locally by chatarreros, people from outside Gigante who buy the product and sell it in neighboring towns. In this sense, artisanal fishing in Gigante produces very little waste and benefits a large number of people since almost all catch is sold, consumed, or given away.

These data confirm that fishing is an important activity for the entire community. Gigante is a fishing-dependent community, and artisanal fishing contributes greatly to food security and the household economy. In addition, fishing benefits a large number of people beyond those who are directly involved with the activity. Thus, changes brought about by tourism growth and restrictions to access to these vital resources could prove catastrophic for many households in the community.

2.2.2 *The Spaces of Fishing in Gigante*

Decades of close-to-shore fishing and local knowledge of site-fidelity behavior shown by some fish species such as snappers, as well as rocky-bottom preference of lobster, have resulted in an in-depth knowledge of local ecology and bathymetry among Gigante fishermen. Local fishermen use this accumulated knowledge and regularly visit the most productive fishing grounds, switching from spot to spot with shifting catch availability. Fishermen classify each spot by what species can be caught and what method should be used. With the exception of Dorado and shark fishing, which take place farther off shore (Dorado is mostly caught beginning at 10 km from the coast), all of these fishing grounds are located close to the coast and overlap with areas of current and future tourism activities, including the marine protected area.

---

commercializing these smaller lobster tails is illegal, the increase in local demand for lobster from tourists gives fishermen an incentive to harvest restricted sizes and acopios to sell them to local restaurants.

The reason that fishermen have recognized the area to be included in the MPA as highly productive and the desire of other actors to preserve the same area can be perhaps explained by a study conducted on coral communities, which showed that the seas around La Anciana Island were the most biodiverse along Nicaragua's Pacific corridor (Alvarado et al. 2011). This biodiversity has had commercial and subsistence value for Gigante's fishermen for generations, but now also aesthetic value for the tourism industry (Fig. 2.3).

In addition to the accumulated knowledge of close-to-shore productive waters, there are also factors that limit fishermen mobility and that have created specific fishing spaces near the coast. Particularly, the small fiberglass pangas used by fishermen in Gigante cannot venture far-off shore during rough ocean conditions. The bad state of many of the engines used, further contributes to this limited mobility. As a result, the limited mobility has created seasonality in the fisheries of Gigante. Between January and April, ocean conditions do not permit fishermen to venture more than 1 to 2 km from the coast. During this time, fishing efforts focus on targeting high-value lobster and snappers in coastal waters. When ocean conditions improve (May to October), fishermen can also begin targeting offshore pelagic species like Dorado and sharks. However, it is not uncommon to read in the newspapers or hear stories about fishermen lost at sea, caught in winds that their small pangas and old motors

Fig. 2.3 A local fisherman weighs his snapper catch at one of the acopios. Monopolization of gasoline and direct commercial ties with buyers forces fishermen to sell their catch to acopios, giving the middlemen a privileged position in the political economy of Gigante's fisheries. *Source* N. A. Alvarado

could not withstand—some are found days later drifting in the currents, others are not so lucky (see Jarquín 2012). It is precisely this lack of reliable boats and motors that make open ocean and multi-day fishing impossible for most local fishermen. Near-shore fishing grounds, then, support local livelihoods.

Access to the shoreline is also important for locals because not all fishing takes place in pangas. Fishermen from Gigante and neighboring communities often prefer to walk to nearby rocky shores either because this is the only access they have to the resource or because it represents a reduction in operating costs (Fig. 2.4). In addition, many local fishermen use the rocky shores when debt with the acopios from gasoline purchases is too high. On a daily basis, the entire shoreline from Punta Pie de Gigante to Manzanillo Beach and even La Anciana Island is dotted with fishermen trying to bring fish home for the day. While using hand line is the most common method for fishing from the shore, some locals swim out to nearby rocky areas to set up a permanent net that they check every early morning for lobster and chatarra. Access to these coastlines is indeed vital for many fishermen in Gigante and the vicinity.

The results of our investigations into the spaces of fishing by Gigante's artisanal fishermen show that, in general, near-to-shore waters are extremely important for local fishermen who in general do not have the equipment to fish in waters beyond 10 km from the shore. Any changes in access to near-to-shore waters will impact local

Fig. 2.4 Access to the shoreline is vital for many fishermen who do not own pangas or who seek to reduce operation costs and debt with acopios. *Source* N. A. Alvarado

livelihoods. Prior to implementation of any regulations regarding access to waters within the boundaries of the MPA, regulators must take into account past and the current use of sea space, such as those illustrated by this study.

2.2.3 Economics of Fishing in Gigante

Fishing is an expensive endeavor for artisanal fishermen and they work hard to stay out of debt. A panga costs at least 4000 USD, and an average outboard motor (60 horsepower) costs 5000 USD. Trasmallos (Fig. 2.5a, b) need to be constantly repaired, which takes days of work, and regularly they need to be replaced all together. A snapper fishing line is 1 km long and made of an average of 500 hooks, with a cost of about 200 USD per line (Fig. 2.5c). Dorado lines are far more expensive (average 500 USD) and thus can only be purchased by acopio owners who then hire fishermen to drive their pangas and use their Dorado longlines (Fig. 2.5d).

Gasoline is by far the biggest ongoing expense at an average of 6 USD/gl. (summer 2012). Near-shore trips to place, check, and recover nets require less gasoline (~3 gallons) and are therefore the most common fishing excursion in Gigante. On the other hand, trips to more distant locations can consume between 10 and 12 gallons. On bad fishing days, fishermen rapidly accumulate debt with the acopios and turn to onshore fishing to try to recover from economic hardships. In fact, it is the key position occupied by acopios as the middlemen in the commercialization of fish in Gigante that contributes to making fishing expensive for fishermen and in keeping fishermen at the bottom of the fishing economy structure.

The acopios work directly with two exporting firms based in Managua (Inversiones Nicafish SA and Central American Fisheries), thus fishermen have no direct commercial ties with these firms. Trucks from these two firms come twice a week to Gigante bringing in gasoline, ice, and fishing supplies like hooks and line. They also bring cash to pay acopios, who then pay fishermen. Acopio owners also sell the gasoline and ice to fishermen and discount the price from the incoming catch. This fish is then taken to processing plants in Managua and prepared for export to the United States and Europe. Because of this direct link between acopios and exporting firms, acopio owners have been able to make handsome profits from the labor of the fishermen. The unavailability of gasoline in Gigante, other than through the acopio, obliges fishermen to sell their catch to the middlemen in order to pay for gas. "He never runs the risk of losing, he always makes money; that's why the fisherman is always at the bottom [of the socio-economic strata]," Igor lamented when asked about their commercial relation with acopios (Igor 2012). Acopios also sell directly to tourists, restaurants, and large-scale tourism developments along the coast.

The expenses related to fishing add yet another aspect that contributes to the limited mobility of fishermen and their reliance on near-to-shore waters. In addition, the current economic structure in Gigante prevents fishermen from improving their

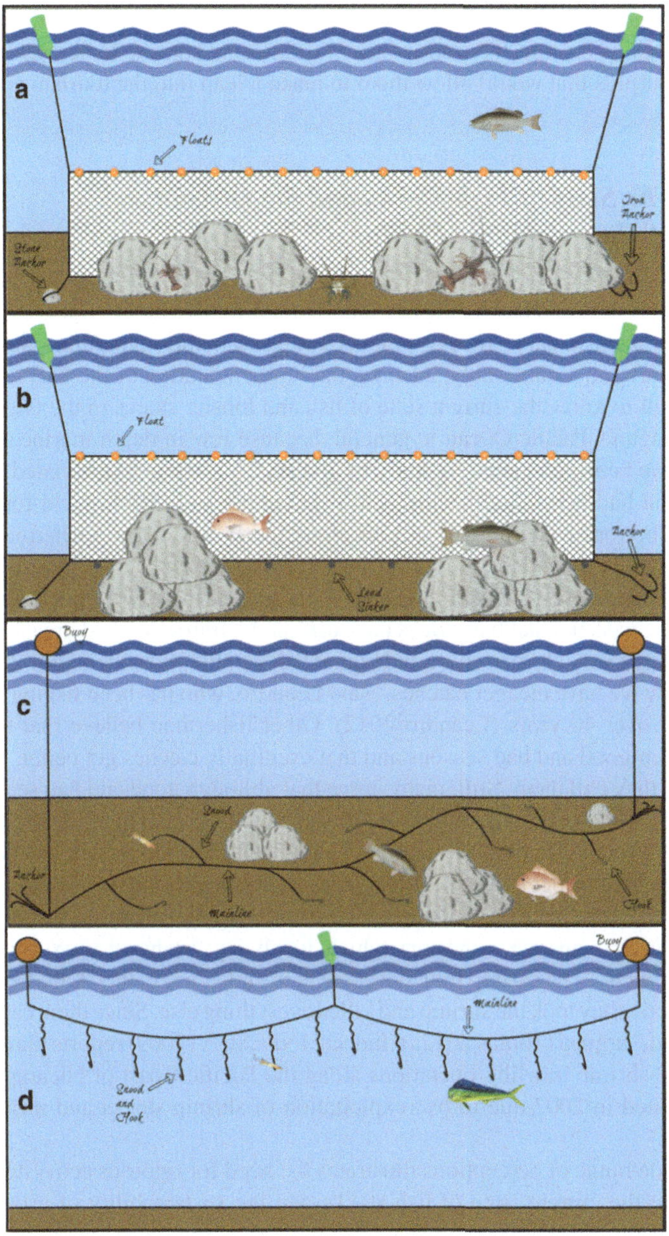

Fig. 2.5 The different fishing methods used by Gigante's fishermen. **a** The trasmallo used to target lobster rests on the ocean bottom and is placed very close to shore in areas with rocky substrate. **b** The net used to target snapper rests just above the ocean bottom and is placed in areas 10–30 m deep. **c** The snapper longline rests on the bottom of the ocean and is placed in areas known to have underwater rocky formations. **d** The longline to target Dorado floats near the ocean surface (10 m max) and is placed in open waters beginning at around 10 km from the coast. *Source* Alvarado and Taylor

condition and allows acopios to accumulate capital that is often used to diversify their sources of income. This structure, in turn, contributes to fishermen's inability to create surplus that would allow them to make a leap into the tourism economy.

2.2.4 *The State of Fish Stocks and the Discourse of Unsustainable Fishing*

Stakeholders in the tourism industry have started to use the discourse of declining stocks and unsustainable fishing practices to presumably undermine the artisanal fishing effort and, conceivably, advance the implementation of the MPA. However, it is difficult to know the current state of fish and lobster stocks in the Gigante area, and Nicaragua's Pacific Ocean in general, because few in-depth marine ecosystem studies have been conducted. A study by INPESCA (2008a, b) estimated that stock exploitation had been significantly below maximum sustainable yield for snappers and other bottom-dwelling fish species along the Pacific coast. Fishermen perceptions about the past and the present resource abundance and declining stocks varied. Fishermen that have fished the longest believed that there was much more abundance in the past and that fish stocks have declined. "Before we had great amounts of sardines come very close to shore and bigger fish would follow them, now they are all gone…now we have meager catches," said Leandro, who has been fishing Gigante's waters for over 40 years (Leandro 2012). Other fishermen believe that there have always been good and bad seasons and that eventually catches get better. "Times of bonanza," they call them. Still, many argue that although good and bad seasons have always been the norm, bad seasons are occurring more often than in the past.

No two groups agree on the reasons for the perceived decline in catches. While government authorities and stakeholders in the tourism industry point to artisanal fisheries and access to open commons, fishermen point to the industrial shrimp trawling ships that used to operate in the area, which, locals say, produced an exorbitant quantity of by-catch. "The shrimp trawlers came here and you see, they killed everything, all the larva…they took the shrimp and killed everything else. Since then, catches have plummeted," argued (Tomás 2012a). Indeed, INPESCA (2009) reports that industrial near-shore shrimp trawling operations along the Pacific Coast of Nicaragua had to be terminated in 2007 due to overexploitation of shrimp stocks and unsustainable practices.

The wide range of perceptions illustrates the need for rigorous ecosystem studies to measure the current state of fish stocks and the sustainability of artisanal fisheries in this area. Eradicating artisanal fisheries assuming a tragedy of the commons discourse could work in detriment of local livelihoods for the purpose of advancing non-extractive and capital-governed interests that puts natural resource use and management out of the hands of locals.

2.2.5 Tourism, Local Jobs, and Changes in Artisanal Fishing

Tourism has brought new job opportunities to Gigante. Younger generations in Gigante are finding jobs in local bars, construction work, and nearby hotels. Many women find employment as housekeepers in tourism projects and second homes along the coast. While young people see a good economic alternative in paid employment in tourism, many have not turned completely away from their parent's fishing traditions. "Although there is good money working in tourism, I keep on leaving my net here every night so that I can bring some food to my family", said Arnoldo as he was gutting a jurel (Yellow Tailed Jackfish: *Carangidae* spp.) on the rocks at the south end of Gigante's bay (Arnoldo 2012). Arnoldo, who is 19 years of age and works as a bartender at one of the local hotels, represents a new generation of fishermen who see a future in tourism development, but believe that fishing is indispensable to their livelihood.

A few job opportunities related to tourism have also opened for local fishermen with pangas. They take surfers to local surf breaks, and provide sport-fishing expeditions. Partnership between local fishermen and foreign investors is the most common way in which these opportunities become available for fishermen. For example, Ernesto related, "I managed to partner with some gringos, …much of the investment for the engines was on their part. They [the gringos] have a website and tourists come here with the entire package paid for. I take them to different spots to surf and fish" (Ernesto 2012). Ernesto is a fisherman who is now dedicated mostly to tourism activities.

Other fishermen who have tried to partake in the tourism boom have seen their aspirations crushed by tourist themselves. Adolfo made this clear, "I have offered tourists to take them out for the entire day for 50 USD, but they say it's too expensive. Then they go to the acopio and see 300 USD as cheap." In Adolfo's perspective, "it is because they perceive us as poor people and think we should content with less" (Adolfo 2012). The connection that acopios have with the tourism industry as the main providers of fish and the capacity they have to accumulate capital from fishing gives them in fact an advantage in capturing business related to ocean-based activities. In addition, acopio boats provide tourists with beer, water, and fishing rods, making the trip more appealing for tourists. As a result, in the words of Adolfo, "the impoverished fisherman cannot make it even if our prices are lower" (Adolfo 2012). At first glance, tourism activities seem a plausible alternative for Gigante's artisanal fishing force, but the reality is that only a few families have the capital to prepare a panga for tourism, including buying more powerful outboard motors and fishing tackle. Tourist perceptions of local fishermen may also impede fishermen from using tourism to diversify their economy. As the tourism economy expands in the community and resources are enclosed to support tourism activities, the majority of fishermen could be severely marginalized and driven further into poverty. Moreover, limited economic resources available to fishermen make the transition to tourism activities unlikely.

2.2.6 La Anciana Marine Protected Area and Conservation Discourse: Tourism Versus Local Livelihoods

Punta Pie de Gigante, Guacalito Beach (immediately south from Manzanillo Beach), and La Anciana Island contain an important abundance and biodiversity of corals and fish (Alvarado et al. 2011). Based on their findings, Alvarado et al. (2011: 1) proposed "management action to protect the diversity and uniqueness of this region." The idea of protecting the area or, as local fishermen see it, prohibiting fishing from Punta Pie de Gigante to Punta Brito, has been floating around in Gigante since at least 2009. Tomás, a local fisherman, explains, "tourism operators proposed to us not to fish close to shore…we said O.K., but that it was necessary to plan for compensation. If they were going to impose a ban on fishing, then we needed some type of monetary indemnification. We live off fishing and they want to prohibit it… how are we going to survive? They said that they were going to see if it was possible to establish some type of compensation but never came back…that was like three or four years ago" (Tomás 2012b). The fishermen see this episode as an idea that never moved forward. Because they have not heard back from anyone regarding alternative plans for their livelihoods and because many of the productive fishing spots are located within the boundaries of the MPA, fishermen state that they would simply ignore any imposed fishing bans in this area.

Management plans for the MPA are not clear at this point. Although the Municipality of Tola officially approved the designation of the area as an MPA in February of 2013 (Alcaldía Municipal de Tola 2013), it is still in the management planning stages and in the hands of two organizations: Fundación Nicaragüense Para el Desarrollo Sostenible (FUNDENIC-SOS) and Fundación Futuro Latinoamericano. These organizations claim (on their website) to be working on a "Proposed MPA with unique mix of corals, rocky reef and sandy bottom, exceptional species richness, close to major turtle nesting beaches on Pacific coast" (Fundación Futuro Latinoamericano 2012). They also state, "fishing communities endure severe poverty" and that "the two adjacent municipalities have about 1,100 fishers in 4 main fishing villages." Finally, they state, "the company planning adjacent tourism resort seeks collaboration with FFI/FUNDENIC." The description reveals two important pieces of information. First, the number of artisanal fishermen that will be affected by this marine protected area is much larger than just Gigante's fishermen, as it will affect livelihoods in four fishing villages along the coast. Second, the description makes explicit that Guacalito de la Isla, a 250 million USD, 1670-acre oceanfront resort for affluent tourists (see Canales Ewest 2013; Felsenthal 2012), requested the implementation of the MPA directly in front of their resort. Guacalito de la Isla is being developed by the Pellas, one of Nicaragua's economically most powerful families and is located directly onshore from La Anciana Island. One of the components of their master plan is a marina from where a variety of marine recreational activities are hosted in the surrounding waters. Tourists fish for snapper and grouper and also dive among coral formations around Anciana Island (Guacalito de la Isla 2012). These activities, however, overlap spatially with the past and current fishing grounds of Gigante's fishermen. Under these circumstances, we must question who will benefit from the establishment of the MPA? At the very least, one needs to question if efforts are truly concentrated on diversifying local livelihoods and protecting marine environments to promote community well-being, or, instead, are conservation discourses being used to conceal capitalistic interests amidst tourism growth in the area?

Guacalito de la Isla see themselves as promoters of the local culture, arguing that their clients, as part of their Nicaraguan experience, want to see and participate in local culture. It is not clear, however, how "local culture" (a fishing community) will be promoted and preserved beyond the recruiting of local fishermen to drive tourism pangas, which is what is taking place currently (Fig. 2.6). It is unlikely that paid employment will help preserve the local artisanal fishing culture, particularly when perceptions of this local culture among tourism stakeholders is far from positive. Representatives from large tourism operators often articulated in interviews with the first author that they don't have any particular plans for fishermen, that fishing communities are very stubborn and difficult to work with, and that unless fishermen organize, developers will not be channeling any help their way. This perspective indicates that some tourism developers have adopted a neocolonial stance of seeing tourism as a way to bring communities out their "impoverished" and "uncivilized" states. Developers are willing to help, but first, fishermen "need to get organized" before the fruits of tourism development are allowed to trickle down. Developers

Fig. 2.6 Young locals hitch a ride on top of an oxcart carrying trasmallos. These types of scenes portraying local culture could come to an end with increasing tourism and conservation discourses that displace fishermen without carrying serious historical, economic, and environmental studies. *Source* N. A. Alvarado

also don't seem to be concerned about Gigante's culture of artisanal fishing beyond the romanticized stories that can be sold to tourists about "the small, bucolic fishing village, which earned its name from the interesting rock formation on its southern end that resembles the foot of a sleeping giant" (Guacalito de la Isla 2012). Tourism operators in the area have not conducted any in-depth studies into the fishing economy from an economic or ecosystem perspective, which could provide valuable information for making development decisions. With regards to La Anciana MPA, for example, stakeholders in the tourism industry, including the central government, have not taken the steps to understand and safeguard resource use patterns among the original users of the coast—the artisanal fishermen of the area.

Our findings provide information on the importance of artisanal fisheries to households in Gigante both for food security and household income-generating purposes, as well as the ocean spaces that are most important for supporting artisanal fishing and the factors that prevent fishermen from moving to different fishing grounds. In addition, findings provide detailed information on the economic barriers that prevent fishermen from moving up in the economic structure within the fisheries industry and thus limit their ability to diversify their livelihood strategies to take full advantage of the tourism boom. We also discussed the perceptions on status of fish stocks and the veracity of unsustainable fishing discourse used by tourism operators to undermine artisanal fisheries. Lastly, the use of conservation discourse to position tourism operators as conservationists and to support the creation of La Anciana MPA could bring about struggles over access to marine resources, especially if the MPA is implemented without considering how resource enclosure could eradicate the fishing culture and economy of Gigante.

2.3 Discussion and Conclusions

The interaction between tourism development and artisanal fisheries in Gigante provides yet another example of how economic and political power is used to achieve the goals of stronger stakeholders in struggles over control of natural resources (Blaikie and Brookfield 1987; Bryant 1992; Neumann 2005; Peet and Watts 2004; Robbins 2012). Gigante will see increased struggles over control of and access to marine resources as tourism starts to dominate local and regional politics, economics, and resource control. Tourism development was first promoted in Nicaragua as a way to bring foreign exchange to an ailing economy and to help alleviate poverty and diversify livelihoods in rural communities. The central government gave little attention, however, to the ways in which tourism could compete with local livelihoods and exacerbate poverty by marginalizing local producers. This is alarming because, in the case of Gigante, artisanal fisheries provide income to local fishermen and an easily accessible food source for many households in Gigante and neighboring communities.

In Gigante tourism has started to compete with artisanal fisheries—an interaction that is beginning to forge new patterns of resource control that favor those with power and money. These new patterns of resource control may end up marginalizing local fishermen and triggering unsustainable resource exploitation (see Robbins 2012). The privileged position of local and foreign elites, despite a brief period of empowerment for the peasantry during the first Sandinista government (1979–1990), once again works to the advantage of the elite—the poor, in this case, are once again marginalized. The Sandinista government (back in power since 2007), paradoxically, stands to see the very people that made the revolution a success, raped by this latest form of capitalist intervention through neoliberal policies that favor tourism development and disempower fishermen. Also, through the adoption of a conservation discourse, the tourism industry has started to label fishermen as rapacious exploiters of the commons, and have promoted themselves as conservationists.

The creation of La Anciana MPA is an example of conservation and control (Robbins 2012). Robbins proposes that control of livelihood-supporting areas is often snatched from local producers under the discourse of environmental and community conservation and by depicting local subsistence practices as unsustainable. This creation of protected areas, Robbins argues, often leads locals who previously managed resources in a sustainable fashion, to break rules and often use the new protected resources in less sustainable ways (2012). On the Pacific coast of Nicaragua, La Anciana MPA will enclose a large and extremely important marine corridor, which includes some of the most important fishing grounds for Gigante's fishermen.

Banning coastal fishing within the MPA may have several outcomes on local livelihoods and environments. First, prohibition of fishing in nearby productive fishing grounds may force local fisherman to fish more distant waters beyond the proposed exclusion zone that stretches up to 22 km offshore. This option, though, is only possible for fishermen who already own better equipment and for acopios. The creation of the MPA will see the complete eradication of on-shore fishing from the rocky shores on the coast and La Anciana Island. This will threaten food security for hundreds of families.

Another outcome is that local fisherman may be forced to employ tactics that get around the rules created by the elite and powerful (cf. James Scott 1990). In what Scott (1985) calls weapons of the weak, fishermen can use everyday resistance by, for example, pilfering inside the boundaries of the MPA, or by fishing at night. This will ultimately allow fishermen to informally complain, while avoiding a full-blown conflict with the powerful stakeholders of the tourism industry, including the government.

Another possible scenario is that local fishermen could use the power and experience gained during the revolution and conflicts over land, to protest their lack of inclusion in the creation of the MPA. The success of MPAs in other parts of the world has been directly linked to community empowerment and the inclusion and adequate consultation of local producers during the planning and implementation stages (Chuenpagdee et al. 2013; Gonzalez and Jentoft 2011; Ho et al. 2012; Kareiva 2006. In addition, from a fisheries management perspective, the implementation of an MPA must germinate from rigorous ecosystem studies that ensure that the proposed ecological functions will take place given the particularities of the marine system (Chuenpagdee et al. 2013; Hilborn et al. 2004; Stevenson et al. 2013).

Unless tourism operations, local and central governments, and institutions create alternatives for fisherman, the implementation of the MPA may result in fishermen displacement and the enclosure of a resource that local families depend upon for food security and income generation. A glimpse of potential future conflicts created by tourism development and the establishment of the MPA is seen in Adolfo's fishing excursion interaction with the sport-fishing boat. Little doubt remains that Adolfo's dilemma reflects the future reality of Gigante's artisanal fishing tradition as it barrels toward collision with the competing interests of tourism development.

Lastly, the power of the large tourism operations and their influence in the Nicaraguan government can subsume local fishermen, prompting them to transition to paid employment in the tourism industry and, potentially, bringing Gigante's artisanal fisheries tradition to an end and with that, the society and culture that the tourism industry attempts to sell. Despite the questions and intent of revolutionaries like Rafael, the ocean seems to be falling into the hands of those with power. Paradoxically, the revolution that empowered peasants in the 1970s to the 1990s now takes a neoliberal turn to create new landscapes where, yes, there are owners of the sea—dueños del mar.

References

Adolfo, fisherman, aged 55, personal interview, December 6, 2012

Alcaldía Municipal de Tola (2013) Ordenanza Municipal que declara y define La Anciana como paisaje terrestre y/o marino protegido. *Ordenanza Municipal* 02–2012

Alvarado JJ, Ayala A, Alvarez del Castillo-Cárdenas A, Alexandra P, Fernández C, Aguirre-Rubí J, Buitrago F, Reyes-Bonilla H (2011) Coral communities of san juan del Sur, pacific nicaragua. Bull Mar Sci 87(1):129–146

Ana, housewife, aged 52, personal interview, Gigante, June 10, 2013

Arnoldo, fisherman and bartender, aged 19, personal interview, June 28, 2012

Badjeck M, Allison EH, Halls AS, Dulvy NK (2010) Impacts of climate variability and change on fishery-based livelihoods. Marine Policy 34:375–383

Béné C (2003) When fishery rhymes with poverty: a first step beyond the old paradigm on poverty in small-scale fisheries. World Dev 31:949–975

Blaikie PM, Brookfield HC (1987). Land degradation and society. London; New York: Methuen

Bryant RL (1992) Political ecology: an emerging research agenda in third-world studies. Political Geogr 11(1):12–36

Canales Ewest G (2013). Resort de clase mundial. La Prensa. (http://www.laprensa.com.ni/2013/01/13/reportajes-especiales/130598-resort-clasemundial). Accessed 23 February 2013

Christie P (2004) Marine protected areas as biological successes and social failures in Southeast Asia. Am Fish Soc Symp 42:155–164

Chuenpagdee R, Pascual-Fernández JJ, Szeliánszky E, Alegret JL, Fraga J, Jentoft S (2013) Marine protected areas: re-thinking their inception. Marine Policy 39:234–240

Cinti A, Shaw W, Cudney-Bueno R, Rojo M (2010) The unintended consequences of formal fisheries policies: social disparities and resource overuse in a major fishing community in the Gulf of California, Mexico. Marine Policy 34:328–339

Derman B, Ferguson A (1995) Human rights, environment, and development: the dispossession of fishing communities. Human Ecology 23(2):125–142

Ernesto, fisherman working in tourism, aged 35, personal interview, July 3, 2012

FAO (2012) *State of world fisheries and aquaculture*, 2012
FAO (2007) Increasing the contribution of small-scale fisheries to poverty alleviation and food security. *FAO Fisheries Technical Paper*, No. 481.125
Felsenthal J (2012) A delicate balance nicaragua. *The New York Times* http://tmagazine.blogs.nytimes.com/2012/11/16/a-delicate-balance. Accessed 23 February 2013
Fundación Futuro Latinoamericano (2012) Gobernanza participativa en áreas marina protegidas de America Central. La Anciana (Nicaragua). https://sites.google.com/a/ffla.net/mpa-america-central/los-sitios/la-anciana-nicaragua. Accessed 23 February 2013
Goldman M (2011) Strangers in their own land: Maasai and Wildlife conservation in Northern Tanzania. Conserv Soc 9(1):65–79
Gonzalez C, Jentoft S (2011) MPA in labor: securing the pearl cays of nicaragua. Environ Manage 47:617–629
Guacalito de la Isla (2012) http://www.guacalitodelaisla.com. Accessed 28 March 2013
Hall R, Edelman M, Borras Jr SM, Scoones I, White B, Wolford W (2015) Resistance, acquiescence or incorporation? An introduction to land grabbing and political reactions 'from below'. J Peasant Studies 42(3-4):467–488. https://doi.org/10.1080/03066150.2015.1036746
Hilborn R, Stokes K, Maguire J, Smith T, Botsford LW, Mangel M, Orensanz J, Parma A, Rice J, Bell J, Cochrane KL, Garcia S, Hall SJ, Kirkwood GP, Sainsbury K, Stefansson G, Walters C (2004) When can marine reserves improve fisheries management? Ocean Coast Manag 47:197–205
Ho TVT, Cottrell A, Valentine P, Woodley S (2012) Perceived barriers to effective multilevel governance of human-natural systems: an analysis of marine protected areas in Vietnam. J Political Ecology 19:17–35
Igor, fisherman, aged 33, personal interview, Gigante, July 12, 2012
INPESCA (2012) *Anuario Pesquero y Acuicultura* 2011. Managua
INPESCA (2009) *Reporte Nacional: Manejo de la Captura y Reducción de Descartes*. Managua
INPESCA (2008) *Estrategia para el Desarrollo sostenible de la Pesca Artesanal, La Seguridad Alimentaria y la Reducción de la Pobreza de las Familias Viculadas 2008–2015*. Managua
INPESCA (2008) *Guía Indicativa. Nicaragua y el Sector Pesquero. Actualización al año 2007*. Managua
Jarquín L (2012) Pescadores de Masachapa perdidos. *El Nuevo Diario*. http://www.elnuevodiario.com.ni/nacionales/271062. Accessed 23 February 2013
Kareiva P (2006) Conservation biology: beyond marine protected areas. Curr Biol 16(14):533–R535
Leandro, retired fisherman, aged 82, personal interview, Gigante, July 14, 2012
Mansfield B (2011) "Modern" industrial fisheries and the crisis of overfishing. In: Peet R, Robbins P, Watts M (eds) Global political ecology. Routledge, New York, pp 84–99
Mansfield (2007) Property, markets, and dispossession: the western alaska community development quota as neoliberalism, Social Justice, Both, and Neither. Antipode *39*(3):479–499
Neumann RP (2005) *Making political ecology*. London; New York: Hodder Arnold
Peet R, Watts M (2004) Liberation ecologies: environment, development, social movements. Routledge, New York
Pennington JT, Mahoney KL, Kuwahara VS, Kolber DD, Calienes R, Chavez FP (2006) Primary production in the eastern tropical pacific: a review. Prog Oceanogr 69:285–317
Pomeroy R, Parks J, Pollnac R, Campson T, Genio E, Marlessy C, Holle E, Pido M, Nissapa A, Boromthanarat S, Thu Hue N (2007) Fish wars: conflict and collaboration in fisheries management in Southeast Asia. Marine Policy 31:645–656
Rafael, fisherman and original cooperative member, aged 56, personal interview, Gigante, June 27, 2012
Robbins P (2012) Political ecology. Wiley, Malden, MA
Scott JC (1990) Domination and the arts of resistance: hidden transcripts. Yale University Press, New Haven
Scott JC (1985) Weapons of the weak: everyday forms of peasant resistance. Yale University Press, New Haven

St. Martin K (2001) Making space for community resource management in fisheries. Ann Assoc Am Geogr *91*(1):122–142
Stevenson TC, Tissot BN, Walsh WJ (2013) Socioeconomic consequences of fishing displacement from marine protected areas in Hawaii. Biol Cons 160:50–58
Stonich SC (2003) Political ecology of marine protected areas: the case of the bay Islands. In: Gössling S (ed) Tourism and development in tropical islands: political ecology perspectives. Edward Elgar, Northhampton, MA, pp 121–147
Tomás, fisherman, aged 58, personal interviews, Gigante, July 16, 2012a; June 20, 2012b
Young E (1999) Balancing conservation with development in small-scale fisheries: is ecotourism an empty promise? Human Ecology 27(4):581–620

Chapter 3
Tourism as Tragedy? Common Pool Problems with Groundwater

There is no more water anywhere, and the poor people are the ones who suffer.
—Personal interview, 23 March 2016

Abstract This chapter examines the increased demand placed on limited water resources by a rapidly growing tourism sector in Playa Gigante, Nicaragua. Results from field campaigns suggest that recharge of the local aquifer may not meet burgeoning tourism demands for water. This chapter also points to initial conflicts over water between locals and tourism operations, which are further complicated by ineffective implementation of national water policies and the common pool nature of groundwater. The conclusion discusses the need for more extensive research and better implementation of water policy through community governance and collaboration.

Keywords Groundwater · Tourism · Nicaragua · Conflict

3.1 Introduction

Nicaragua has long been regarded as the "land of lakes and volcanoes." This imagery conveys a colloquial sense of water abundance, validated in a technical assessment of the water resources of Nicaragua by the US Army Corp of Engineers in 2001. The report concluded that Nicaragua is "rich in hydrologic resources" and that the presence of abundant water resources and rainfall ensure "there is adequate water to meet the water demands" of the country (Webster et al. 2001). Although water rich, Nicaragua has historically faced issues of adequate access to clean water for its people. One of the major problems in water management identified by the report was the absence of a national water law. This potential obstacle to water security was ostensibly addressed in 2007 with the passing of Ley General de Aquas Nacionales (General Water Law). Law 620 defined water as a public good and framed governance so as to ensure sustainable and equitable use of water and the promotion of social and

G. T. LaVanchy et al., *Tourism in Post-revolutionary Nicaragua*,
SpringerBriefs in Latin American Studies,
https://doi.org/10.1007/978-3-030-55632-7_3

economic development. Within this framework, water was prioritized among uses and users when granting concession, permits, or license for water use.

Despite a perceived abundance of water resources and robust supporting law, water has become a point of conflict in southwest Nicaragua. This conflict is derived from a variety of physical and social issues and has resulted in hardship for nearly all stakeholders. The objective of this chapter is to explore the linkages between the growth of tourism, groundwater quantities, and local access to water in southwest Nicaragua. We use findings from semi-structured interviews, groundwater and meteorological monitoring, and geological mapping to explain the physical and human geography contributing to conflict between stakeholders with various levels of knowledge, power, and access. The chapter is divided into four sections. First, the connection between tourism and water is traced from a global scale to the local context of Gigante. Second, the geography of water in southwest Nicaragua is outlined, with particular attention given to precipitation and geological conditions specific to the coast. Third, the water struggle and its consequences for local population and tourism developers are described. Social power is scrutinized through examples of competition over limited water supplies of water, while economic consequences for tourism developers are projected. Lastly, implications and recommendations for water management are offered.

3.2 Water and Tourism

As pointed out in Chap. 1, tourism is often promoted as a means for driving economic growth and providing jobs to local economies. A counter argument that emphasizes the capacity of tourism to produce deleterious social and environmental outcomes is particularly salient with respect to water. The material dependency of tourism on water has been well documented by Gössling et al. (2015) and others (Essex et al. 2010; Hadjikakou et al. 2013; Rico-Amoros et al. 2009). Although agriculture accounts for the largest demand on global freshwater, tourism is a major driver of water consumption, often in dry destinations with limited water resources. Research shows that tourists are likely to use substantially greater amounts of water than local residents, in some cases up to 15 times more (Gössling 2001). In part, this is due to the water demands of landscaping, swimming pools, and golf courses that accompany luxury-type tourism in dry landscapes.

It is not surprising that tourism found the municipio of Tola and community of Gigante. Indeed, this area is similar in appearance and attractiveness to the northern reaches of its southern neighbor Costa Rica, yet has only a fraction of the population density and tourism pressure. Like Costa Rica, the southwest coast of Nicaragua is a paradise for surfers due to consistent, abundant waves. Low population density and affordable land make it particularly suitable to residential-type tourism and gated communities—a type of luxury tourism that yields more revenue per tourist visit, but is heavily dependent on water. Numerous studies on the environmental impacts of tourism on host countries are relevant for emerging tourism destinations such as

Nicaragua where data are often lacking. Several papers examined the comparative water demand of "quality" and "mass" tourism, where quality tourism is characterized by low density, second homes and golf courses, and mass tourism by intensive vertical hotel growth, and high season sun and sand consumption. Hof and Schmitt (2011) and others (Deyà Tortella and Tirado 2011; Rico-Amoros et al. 2009) showed that water consumption patterns are actually higher for low density, quality tourism than mass, or hotel tourism. In part, this is due to the gardens, swimming pools, and golf courses that accompany quality tourism. Although tourism is a highly differentiated activity, these findings are significant to the growth of tourism in southwest Nicaragua that is trending toward low-density models. It is worth noting that the water demands of similar type tourism in northern Costa Rica has also produced conflicts among "local" and "outside" stakeholders (Kuzdas et al. 2016).

3.2.1 Water in Southwest Nicaragua

Nicaragua is home to three Köppen-Geiger climate types including fully humid equatorial (Af), monsoonal equatorial (Am), and winter dry equatorial (Aw). The area of Gigante belongs to the latter category, also known as dry tropical forest, and is characterized by unevenly distributed rainfall and distinct wet and dry seasons. Nearly all of the average rainfall of 1450 mm occurs May to October, with the canícula (brief summer drought) breaking the rainy season in July (Fig. 3.1).

Given the lack of surface water in the study area, groundwater is the only option for meaningful quantities of freshwater. The amount of groundwater available to

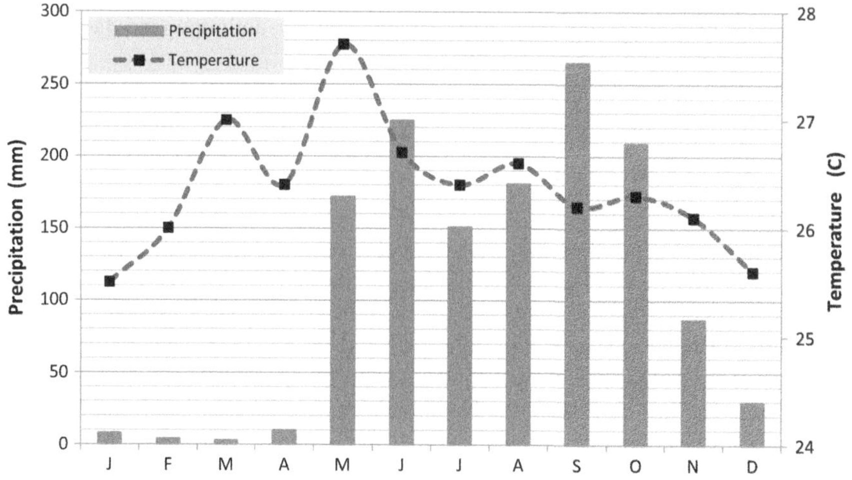

Fig. 3.1 Climograph for Tola region. *Source* G. T. LaVanchy and Instituto Nicaragüense de Estudios Territoriales (INETER)

local residents and tourism development is thus directly linked to the amount and timing of rainfall—two factors in facilitating groundwater recharge.

3.2.2 Geology

Despite a modest amount of annual rainfall, recharge to the aquifer is complicated due to geological conditions. Gigante lies in a watershed underlain by the Brito geological formation, a 2500 m thick sedimentary sequence of shales, limestones, sandstones, siltstones, and mudstones, with pockets of volcanic breccias and tuffs (Arengi and Hodgson 2000). The upper extent is mostly fine-grained sandstone with negligible porosity of 2% or less (LaVanchy 2017). Thus, groundwater movement within the aquifer is largely controlled by bedding planes and vertical fractures (secondary porosity), making well productivity spatially variable and challenging to predict. Rainfall infiltration (i.e. groundwater recharge) is also constrained by the low permeability of surface rocks. Practically, this means that heavier or more intense rainfall events result in excessive runoff to the ocean rather than meaningful contributions to aquifer recharge. Collectively, these are problematic to aquifer yield and serve to complicate the process of siting wells and predicting output.

Most of the wells in Gigante are manually dug (artisanal) and extend from 5 to 15 m below the surface to intersect the water table (Fig. 3.2). These generally produce limited quantities of water (between 0.5 and 5 L/min) and are typically shared between several households. Drilled wells are few in number and are owned by larger tourism developments or resident foreigners. These deeper wells (35–260 m) typically produce greater volumes (75–450 L/min), though they are susceptible to saltwater intrusion due to their depth, proximity to the ocean, and often high rates of pumping. In addition to rainfall and regional geology, anthropogenic factors also play a critical role in water availability.

3.2.3 Water Law and Common Pool Issues

Water management around the world can be categorized by two end-member approaches—free market and government command and control. The latter approach is employed in Nicaragua through a robust and nuanced national water law (Law 620 of 2007) that established water as a public good and provided a framework for the state to ensure its role in social and environmental well-being and to protect against over-abstraction. The law empowers the National Water Authority (Autoridad Nacional del Agua, ANA) to manage national water resources and regulate allocation of water rights among users. In part, ANA's mandate is to "prepare a water balance for each basin" and "propose management regulations for basins and aquifers" (Republic of Nicaragua 2007, 20). However, neither of these directives have yet to be implemented

Fig. 3.2 Typical hand-dug (artisanal) well in Gigante. *Source* G. T. LaVanchy

for any of the nation's 21 water basins, thus rendering any notion of recharge quantities or safe-yield extractions for tourism development untenable. Without guiding data or government regulatory oversight, tourism developers are merely poking straws in the ground until they find sufficient quantities of water to suit perceived needs. This uninformed manner of water abstraction is unstable for tourism developers from both a fiscal and water security standpoint. Drilling unproductive wells is costly, while over pumping existing wells can lead to saltwater intrusion. The absence of sustainable water management from the tourism industry then puts local populations at risk as declining water tables fall below their capacity to manually dig wells (hand dug wells in this area are typically no more than 15 m deep due to the physical challenges incurred by excavators).

Groundwater functions as a common pool resource, subject to competition among users. Broadly speaking, "common pool resources" (CPRs) are shared resources in which each stakeholder has an equal interest and whose boundaries are difficult to delimit. Consequently, exclusion of users of the resource is difficult (referred to as non-excludability) and consumption by one individual reduces the amount available for other consumers (referred to as subtractability) (Ostrom 1990; Bromley 1991). Resources under these conditions are often associated with the "Tragedy of the Commons" outcome, promulgated by ecologist Garrett Hardin in 1968. His thesis stated that resources held in common (i.e. land, ocean, air, and rivers) are inevitably subjected to the tragedy of overexploitation given that individual rationale generally prevails over collective rationale when commonly held resources are managed. He reasoned that individuals would maximize their use of the resource at the expense of the community, thereby exceeding carrying capacity and bringing the commons to ruin (Hardin 1968). To avoid this outcome, Hardin concluded that the commons should be either privatized or managed as a public property via allocation of use and right to entry. These two solutions represent two ends of a continuum for dealing with limited resources, the latter providing the conceptual framework for Nicaragua's national water law (N° 620). However, the absence of hydrological knowledge and enforced water regulations creates a de facto tragedy of the commons over groundwater in southwest Nicaragua.

Briassoulis (2002) and others (Cabral and Aliño 2011; Healy 1994) have examined the role of common pool resources in sustainable tourism. Fundamentally, tourism is characterized as an activity which consumes natural, socio-cultural, and built attraction "resources" in host areas, often in competition with local populations (Briassoulis 2002). However, most of the literature on the tourism-CPR nexus deals with natural landscapes, facilities, and cultural amenities that experience "degrading" from overuse. Water is only considered in broad, conceptual terms, with no assessment of the contribution of non-excludability and subtractability to conflict. Groundwater has been analyzed as a CPR outside the field of tourism studies, most notably by Brentwood and Robar (2004). Their edited volume took an international look at the contributions of groundwater to socio-economic development and concluded that the common pool nature of groundwater "represents a significant social-management threat" (xii). They argued for a two-fold strategy to avoid a tragedy of the commons. National governments must "limit the number of users, or regulate the amount each user can take from the commons" (xii). Additionally, conservation of groundwater must be prioritized and encouraged through policy. Returning to the issue of the national water law of Nicaragua, LaVanchy et al. (2017) noted that Law 620 has great potential to effectively manage water, but as of yet has no practical success of solving water conflicts due to the barriers of information gathering, property rights protection and enforcement, and strategic costs. Thus, water conflicts in Gigante and other areas undergoing rapid tourism development along the Pacific coast reflect a gap between policy intent and successful water management. Left unattended, this gap translates to deleterious outcomes for all stakeholders.

3.3 Evidence of Conflict

The sharp rise of tourism in the municipality of Tola in recent years has led to unpredicted pressures on water resources. Tourism developers use varying amounts of water to provide services to customers, including cleaning food (restaurants), cleaning boats (fishing charters), laundry service, showers, swimming pools, landscaping, and keeping golf courses green (to name a few). Additionally, an increase in population in Gigante due to in-migration for tourism-related employment also

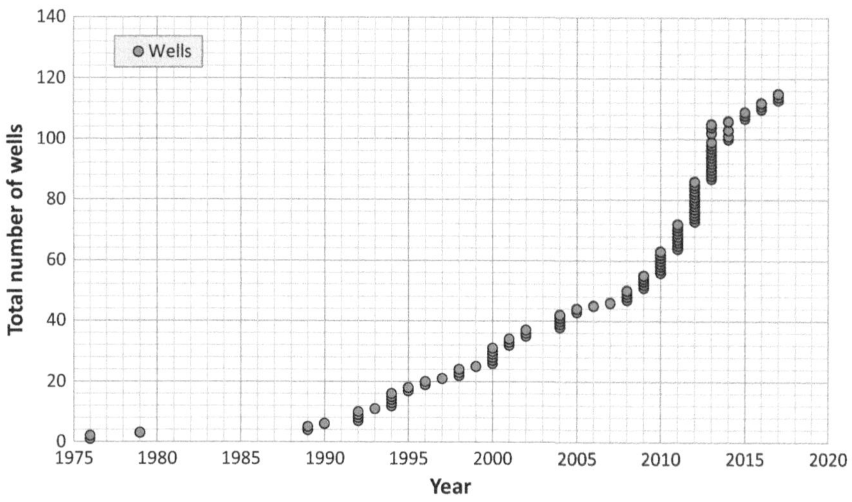

Fig. 3.3 Proliferation of wells in Gigante. *Source* G. T. LaVanchy

contributes to the collective increase in demand for water. This growth in demand can be "seen" through the dramatic increase in number of wells in the area since the mid-1990s (Fig. 3.3). This proliferation of wells means that more "straws" are pulling from the aquifer(s).

The dramatic growth in number of wells and subsequent water abstraction is compounded by varying climate (i.e. cyclical droughts) and predicted climate change. Given that the southwest coast of Nicaragua has minimal surface water resources, tourism development is particularly dependent on groundwater (replenished through rainfall) to meet its consumptive needs. However, this region is predicted by Global Circulation Models to experience drying conditions, which in turn will impact aquifer recharge and subsequent water quantities available for abstraction. The Intergovernmental Panel on Climate Change (IPCC) Fourth Assessment Report (AR4) reported that Central America is "likely" (>66% probability) to experience declines in rainfall in both dry and wet seasons, and Nicaragua specifically will see an increase in frequency of dry days (IPCC 2007). These predictions are problematic for the current trend of tourism growth along the southwest coast of Nicaragua given the linkage between precipitation and aquifer recharge, on one hand, and tourism and groundwater abstraction on the other.

Rainfall has fallen short of the long-term annual mean of 1450 mm for the past five consecutive years (Fig. 3.4). Although the 49-year rainfall record reveals periodicity of drought, previous events did not coincide with tourism growth and its accompanying increase in groundwater abstraction. It is important to emphasize that most agriculture in this region is rainfed and thus not in competition with tourism over groundwater supplies. Over the period 2012–2016, annual rainfall yielded (respectively) 80, 83, 56, 39, and 66% of the annual mean and resulted in the drying of a majority of wells (LaVanchy 2017).

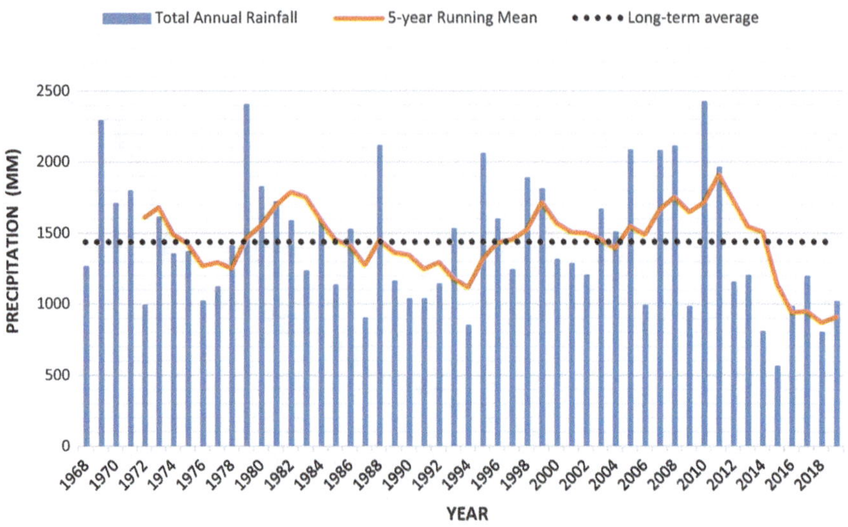

Fig. 3.4 Annual rainfall for study area. Dashed horizontal line represents the mean of 1440 mm for the climatic series of 1968–2019. *Source* G. T. LaVanchy, M. J. Taylor, and Instituto Nicaragüense de Estudios Territoriales (INETER)

The decline in rainfall is characteristic of a region-wide decline that was especially notable in 2015 and 2016 when much of the Pacific Coast region of Central America was affected by insufficient and erratic rainfall associated with a strong El Niño event. Drought conditions reduced crop production for small-scale producers of maize and beans in several areas of Central America between 50 and 100% (Pons et al. 2017). This same drought resulted in a significant drop in static water levels of the aquifers in our research area. The combined influence of drought and tourism related abstraction of groundwater meant that shallow, hand-dug wells (used by 99% of the local population) no longer reached the water table. Over 40% of these wells went dry, while an additional 43% were at a critical stage of less than 1 m of remaining water depth (Fig. 3.5).

Data loggers deployed for continuous measurement in two wells also demonstrated a decline in the water table after the terminus of the 2013 rainy season (Fig. 3.6). The average negative change of 4 m reflects diminished rainfall and increased groundwater abstraction by tourism developers. Interviews with well owners ($n = 90$) revealed unprecedented experiences—none had previously witnessed so many dry wells and nearly 80% acknowledged a growing sense of crisis over water.

Despite a lack of recharge to aquifer(s) in the region and subsequent lowered water tables, tourism developers continue to pump groundwater to meet their needs. Their collective abstraction exceeds recharge rates and has resulted in drawing the coastal freshwater/saltwater interface landward, thus contaminating many of the coastal wells and further contributing to water insecurity. Central to the issues of

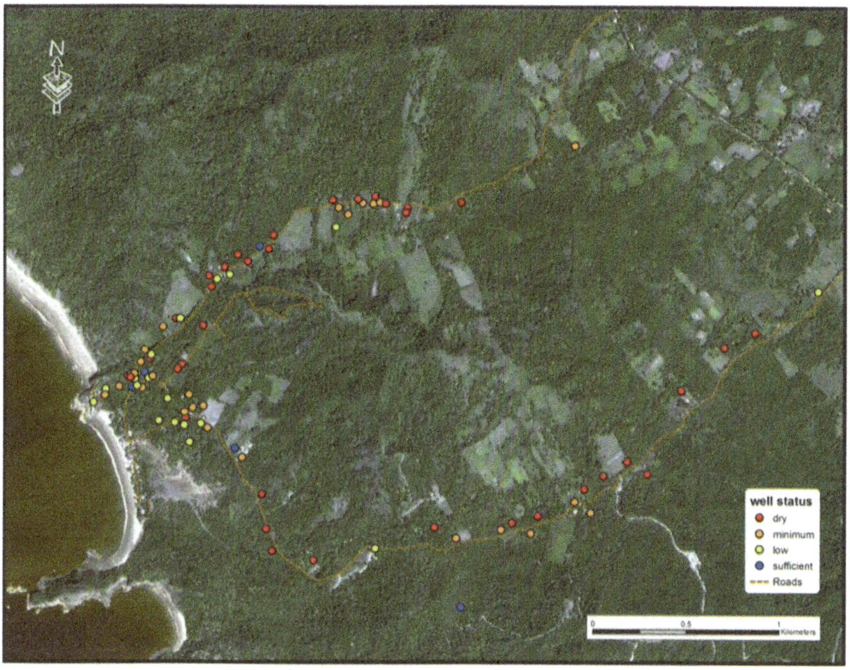

Fig. 3.5 Well inventory from June 2015 revealing the relative quantity of water in Gigante wells. Minimum indicates less than 0.5 m remaining; Low indicates less than 1.0 m remaining. *Source* G. T. LaVanchy and Digital Globe imagery

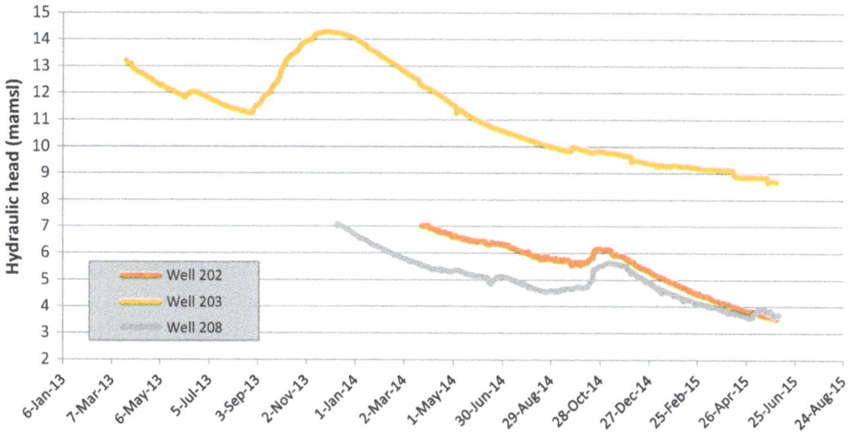

Fig. 3.6 Daily hydraulic head values for two drilled well in Gigante showing an ~4 m drop in the water table from 2013 to 2015. *Source* G. T. LaVanchy and M. J. Taylor

Fig. 3.7 Transporting water from a distant well to home. *Source* M. J. Taylor

drying wells and saltwater intrusion is the general lack of hydrological knowledge by tourism developers. In principal, national water laws should inform and provide oversight of groundwater abstraction. In reality, tourism developers have pumped water for their development needs without knowing the sustainable yield of the provisioning aquifer(s). Beyond endangering the economic sustainability of tourism, this lack of knowledge has led to depleted or saline wells for local populations—a particular hardship imposed by the combined forces of drought and tourism development. As one resident described it, "there is no more water anywhere, and the poor people are the ones who suffer."[1] This hardship takes on discrete significance in national contexts like Nicaragua's (and many other Latin American countries) where responsibilities for rural water management are shouldered primarily by residents themselves, albeit oftentimes in coordination with state agencies and officials (Romano 2016; Dobbin and Sarathy 2015; Llano-Arias 2015; Schouten and Moriarity 2008). In Gigante, residents with dry wells must now secure their water needs by hauling water from other sources (Fig. 3.7).

In light of the common property nature of groundwater in Gigante, the term water "grabbing" (Mehta et al. 2012) is appropriate to describe the activity of securing water amidst prolonged drought and drying wells. As noted of CPRs, consumption by one

[1]Personal interview, March 23, 2015.

individual reduces the amount available for other consumers, and it is impossible, or very costly, to exclude additional users. Given that no water budget or safe-yield has yet been established by ANA for the water basin that provisions Gigante, nothing guides or constrains actors from grabbing water they deem necessary. Several examples are worth noting to illustrate the asymmetrical power relationship between locals and tourism developers, and subsequent marginalizing effect of water grabbing.

Until recently, one of Gigante's two public wells adequately supplied residents with cleaning and cooking water for the entire 30-year history of the town. Laundry was cleaned at the well and water was drawn and carried by hand to nearby homes. In 2013, a tourism operator began pumping and hauling 10,000 L per day to clean charter-fishing boats. Within two months the once reliable public well dried up and remains unusable to date. The tourism operator then began pumping from the remaining public well. After several months of similar abstraction, owners of nearby wells reported increased salinity in their family wells from seawater intrusion. Presently, six of these wells are unusable for potable water sources (Fig. 3.8).

Some forms of water grabbing are more obscure, though equally impactful on water security. In two separate instances, small tourism developers (scrambling to secure their own water needs) purchased small parcels of land adjacent to historically productive wells for the purpose of siting new wells. One of the developers dug a 28 m artisanal well, largely with the aid of expensive air hammer tools unavailable to

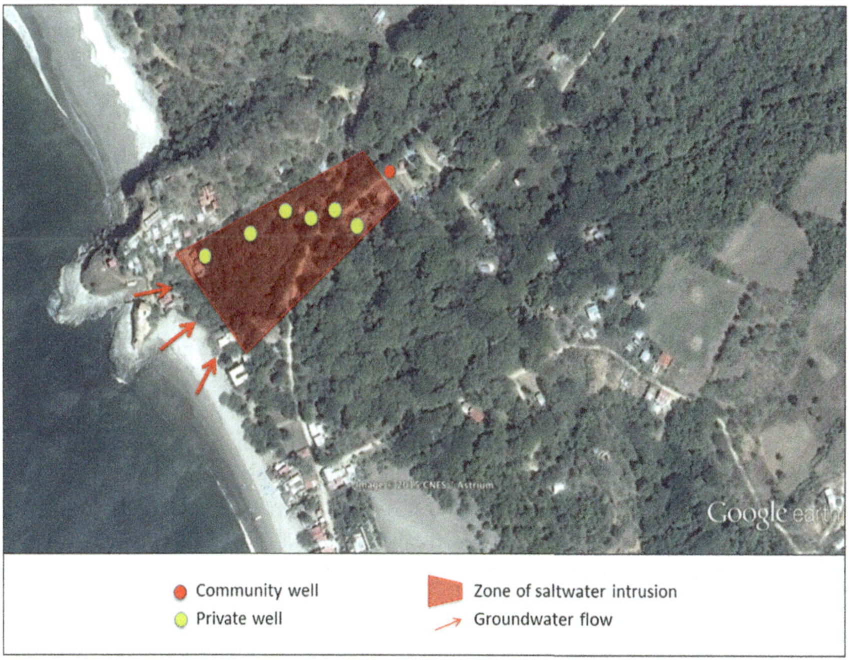

Fig. 3.8 Private artisanal wells contaminated from saltwater intrusion. *Source* G. T. LaVanchy and Google Earth

locals. Shortly after the onset of pumping from the well, adjacent users complained to the tourism developer that their wells were dry. In describing the incident, the developer's initial (and limited) reaction was merely "I felt like I stole their water."[2] Again, the lack of understanding of groundwater movement by tourism developers and the absence of water budgets from ANA creates a scramble for water resources that discriminates against those lacking power or economic means.

Further to this point, residents living near a golf course installed in 2011 reported that once reliable family wells are now perennially dry. One resident reported that he approached the luxury resort owning the golf course to request delivery of water to compensate for his dry well. When they refused, he threatened to take the story to a prominent newspaper. The prospect of negative media exposure forced the resort to capitulate and now they deliver water to all proximate owners of dry wells. While it is difficult to disentangle the impacts of tourism abstraction from the current hydrological drought, the estimated dry season requirement of 1.7 million liters per day for the golf course (QUENCA 2011) undoubtedly plays a contributing roll in the drying of artisanal wells. In the eyes of locals, it is THE cause of their dry wells.

3.4 Discussion

As noted previously, Hardin (1968) argued that common pool resources such as groundwater will inevitably be overused without privatization of the resource or public management via allocation and right of entry. Law 620 of Nicaragua encapsulates the latter category and represents a defining stance on the part of the government to guard against a winner/loser outcome with respect to invaluable resources such as water. The law accomplishes this in principle by designating water as a national resource under the managing responsibility of the state. Further, it requires that water basin budgets be generated for each water basin in the country, and that basin committees be formed to oversee sustainable allocation of water to the various uses and users within the basin. However, groundwater quantities and their spatial distribution are largely unknown in the Gigante region due to insufficient data and research. Without these data, the state is unable to sustainably allocate water and guard against overabstraction. This absence of effective water management leaves critical groundwater resources at the disposal of anyone with sufficient means to extract them. In this sense, Gigante is a bit like the "Wild West" in that almost anything goes with regard to developing tourism and securing the water necessary to support such tourism development. Like other forms of common pool resources, extracted groundwater impacts all other users of the aquifer in the watershed.

Local residents of Gigante worry about their water security as levels in their wells drop precipitously or wells dry altogether. Although most local residents seem aware of the growing demand for water and its limited supply, few informants offered community-wide solutions. When wells are seasonally dry, owners

[2]Personal interview, June 3, 2015.

either excavate additional depth from the well (if possible), manually haul water from further distances, or decrease consumption. Typically, the burden of manually hauling of water falls on the shoulders of women and children, while the expensive cost of deepening of wells upon the entire family. During the 2015 dry season, most locals were forced to deepen wells, paying local well diggers US$20 a day for their work. Typically, a deepened well requires two diggers working over two days at a cost of US$80 to the well owner. This represents a significant financial strain on local households given the average monthly salary of US$160–$200 per month. Alternatively, decreasing consumption of water affects overall hygiene and translates to increased health risks.

On the part of tourism developers, lack of hydrological knowledge and subsequently depleted aquifer(s) and saline intrusion puts them in tenuous economic positions. Most developments were designed with nominal thought to sustainable water supply. The general assumption by developers was that this region had plenty of groundwater to support whatever tourism ensued. At the onset of tourism growth in the early to mid-2000s, tourism operations were small and had less water demands. Since that time, tourism has grown both in number of operations and in size. Too, the bulk of development in coastal southwest Nicaragua consists of gated residential communities—a type of luxury tourism heavily dependent on water. Thus, the issue of declining water availability undermines promotional imagery of the Tola region as the "Emerald Coast." The term is a recent branding effort by prominent tourism developers in order to attract tourists and promote investment. In reality, this coastal area consists of dry tropical forest, characterized by distinct wet and dry seasons (unimodal precipitation pattern) and brown landscapes (leaf-off) for five months of the year. Thus, expectations of verdant landscapes by visiting tourists has "forced" tourism developers to use large amounts of water (up to 65% of total daily usage) to irrigate landscaping for homes and public spaces within gated tourism developments during the dry season (LaVanchy and Taylor 2015).

3.4.1 Whither Tourism?

Nicaragua has moderately held to its revolutionary roots in which the state and the economy are ideologically meant to serve the majority population through a socialist-oriented system. However, in an effort to dislodge its status as the second poorest nation in the western hemisphere, Nicaragua's government has embraced neoliberal economic agendas, including the promotion of tourism. Using tourism as a tool for development is a promising piece of this solution, however, it is unclear whether this tool will provide sustainable socio-economic development or merely perpetuate inequality in wealth. It also remains unclear what impact the tourism agenda will have on water—a critical resource for the country and for tourism development. It is already evident that tourism is impacting water resources in the Gigante area. However, enthusiasm for tourism development remains strong, as evidenced by ongoing construction of boutique hotels, restaurants, and new residences within gated communities. Recent interviews with several tourism developers revealed plans to expand operations, which will inevitably lead to further demand for water. Concomitantly, these developers are actively looking for ways to improve well productivity and for additional areas to drill new wells as they wonder if they will have sufficient water resources to maintain their business. This pressure on groundwater is anticipated to increase given the recent prominent advertising for Nicaragua's Pacific coast in surf, outdoor, golfing, and inflight magazines (Dobson 2015; Dear 2013; Klinger 2014). This increased visibility will likely translate to increased tourism visits and an array of interest in adding more tourism development.

While it is clear that the groundwater commons in Gigante is being overexploited and lacks sufficient oversight, the future of this resource, and subsequently the well-being of local populations and the tourism industry, couldn't be less clear. The meteorological record suggests that 3–5 year drought events are normative; thus, stakeholders must adapt their expectations for abundant water to cycles of dry and wet periods. Further, the national government has been unable to enforce its own water law since codification more than a decade ago, and is unlikely to do so in the near future. (See Herrera 2014 for a general report on the stance of the government towards water sustainability.) A logical question then arises, *will the converging forces of climate variability, absence of government oversight, and tourism driven abstraction of groundwater precipitate a scramble to the bottom where all stakeholders lose, or will actors in Gigante initiate grassroots solutions to water insecurity?*

3.4.2 Paths Forward

Several possible paths toward water security exist for the stakeholders in Gigante, some of which are already in process. Some of the local residents of Gigante have tackled the growing issue of water insecurity through the organization of a Potable Water and Sanitation Committee (CAPS). Local water committees (CAPS) are recognized and encouraged by the central government as a way for rural communities, through collaboration between government and international institutions, to manage their own water resources. Presently, there are thousands of CAPS operating throughout Nicaragua at a variety of levels of success. As might be expected, any success is ultimately a product of the organizational and leadership strength of the local CAPS (Romano 2019). The Dublin Principles contends that successful water management systems entail meaningful stakeholder participation and some degree of government oversight and support. These could be on more formal levels with advisory boards and associations, or scaled-down applications at appropriate levels (Solanes and Gonzalez-Villarreal 1999). To date though, the CAPS in Gigante has not been able to move from the organization to action phase, leaving a question of its efficacy and ultimately the plight of water security in Gigante (LaVanchy et al. 2017).

Some tourism developers are also taking steps to alleviate the crisis via conservation measures. In part, this has taken the form of small placards and signs in restaurants or hostels to remind tourists to save water. The largest conservation impact has been realized through installing water meters on individual houses in most of the gated communities around Gigante. This strategy has brought much needed realization to homeowners of the amount of water being used (often on landscaping), as well as incentivized the pinpointing and remediation of leaking pipes. Several of the gated communities have implemented tiered tariffs, thereby economically discouraging heavier demands for water. To further decrease their water footprint, one gated community replanted water intensive landscaping with plants that use less water, yet still provide color and variation to the seasonally brown hillsides. Only one development in this study indicated a willingness to invest financial resources for a proper hydrological study to determine the balance between their water demands and aquifer water supply.

Each of the measures mentioned above is necessary for productive and equitable tourism, yet can often be costly and do not exclude others from disproportionate usage. An additional barrier to investing in conservation and/or hydrological studies is the disconnect between water managers (who are on the ground) at tourism developments and the executive managers of those developments. The latter are often absent from the property, have little to no knowledge of water management principles, and tend to make short sighted, real-estate-driven decisions. Each of the larger developments we interviewed shared this challenge. Water managers were constantly working to educate their managers, yet were often left to manage the most vital resource with meager financial support. In all likelihood, it will necessitate further drought, salinization of wells, and local hardship to force executive managers of developments to invest adequate finances into hydrological studies.

3.5 Conclusions and Recommendations

This study, using data from interviews with local stakeholders, on seasonal fluctuations in well water levels, precipitation, and geology shows that groundwater is increasingly seen by all actors as a high demand resource with uncertain supply. Moreover, the national water law was designed to promote well-being and equity, but is dependent upon hydrological knowledge of groundwater recharge to guide a "safe-yield" for water extraction by tourism development. The ensuing tragedy of the commons has deleterious effects on all actors. Without investment in research and subsequent knowledge dissemination, conflict over water is likely to increase.

Based on the findings of this study, the following recommendations are made:

- To establish a unified monitoring network of drilled wells in order to allow continuous monitoring of water level changes with respect to recharge and withdrawal. This would require installation of data loggers (for water levels, water temperature, and electrical conductivity) and a central data collection system. Tourism developments should take the lead, but contribution should also be made from single well owners.

- To collect monthly water level measurements for all artisanal wells.
- To carry out additional hydrogeological research to determine aquifer(s) characteristics such as recharge, infiltration, hydraulic conductivities, and fracture network parameters. This would necessitate boreholes and tracer tests to study the geometry of the aquifer. (This is fiscally possible with partnerships between developers, Nicaraguan academics, and the government.)
- To generate a numeric model of the aquifer in order to (1) develop water extraction strategies, (2) estimate cost of extraction, and (3) establish sustainable pumping rates to avoid over-drafting and/or saltwater intrusion.
- To organize a workshop for sharing best practices on effectively tracking water consumption rates.
- To formalize the CAPS in Gigante in order to fairly represent all stakeholders and streamline reporting to the National Water Authority (ANA). Moreover, a well-informed CAPS can hold open conversation among all water users about demand, supply, and fair governance of water to avoid potential conflict.

Water resources are essential to the support of life and livelihood, yet the challenges of supply and demand for water persist in both undeveloped and developed regions of the world. Many countries throughout Latin American are faced with rising pressures of water provisioning due to population growth, increasing urbanization, tourism, and climate change (OECD 2012; Van Noorloos 2011; World Bank 1998). The ensuing water stress and scarcity is complicated by policy strategies, government implementation, privatization, and economic recession. Over the past decade, several countries in Central America have re-structured national laws to define groundwater as a public good in order to ensure equity and sustainability. Although these laws provide citizens with a constitutional guarantee of equitable access to water, it remains unclear if this "commons" strategy will fulfill its intended outcomes. This chapter examined the increasing demand placed on limited water resources by a rapidly growing tourism sector in the Gigante area. It argued that the production of tourism in Gigante is (1) a product of national interest in tourism and a world-class coastal and surf setting, (2) is dependent upon the common property regime of groundwater, and (3) is at odds with environmental constraints (precipitation and geology) that determine available water resources.

References

Arengi J, Hodgson G (2000) Overview of the geology and mineral industry of Nicaragua. Int Geol Rev 42(1):45–63. https://doi.org/10.1080/00206810009465069

Brentwood M, Robar S (2004) Preface. In: Brentwood M, Robar S (eds) Managing common pool groundwater resources: an international perspective. Praeger Publishers, Westport, CT, p xii

Briassoulis H (2002) Sustainable tourism and the question of the commons. Ann Tour Res 29(4):1065–1085

Bromley D (1991) Environment and economy. Blackwell Publishers, Cambridge
Cabral RB, Aliño PM (2011) Transition from common to private coasts: consequence of privatization of the coastal commons. Ocean Coast Manag 54:66–74. https://doi.org/10.1016/j.ocecoaman.2010.10.023
Dear T (2013) Nirvana in Nicaragua, June 2013. Colorado AvidGolfer, pp 28–36
Deyà Tortella B, Tirado D (2011) Hotel water consumption at a seasonal mass tourist destination. The case of the island of Mallorca. J Environ Manage 92:2568–2579. https://doi.org/10.1016/j.jenvman.2011.05.024
Dobbin KB, Sarathy B (2015) Solving rural water exclusion: challenges and limits to co-management in Cost Rica. Soc Nat Resourc 28:388–404. https://doi.org/10.1080/08941920.2014.948245
Dobson J (2015) Nicaragua, the undiscovered Luxury beach destination for billionaires and celebrities, 30 October 2015. Forbes. http://www.forbes.com/sites/jimdobson/2015/10/30/nicaragua-the-undiscovered-luxury-beach-destination-for-billionaires-and-celebrities
Essex S, Kent M, Newnham R (2010) Tourism development in Mallorca: is water supply a constraint? J Sustain Tour 12(1):4–28. https://doi.org/10.1080/09669580408667222
Gössling S (2001) The consequences of tourism for sustainable water use in a tropical island: Zanzibar, Tanzania. J Environ Manage 61:179–191. https://doi.org/10.1060/jema.2000.0403
Gössling S, Hall CM, Scott D (2015) Tourism and water. Channel View Publications, Bristol, UK
Hadjikakou M, Chenoweth J, Miller G (2013) Estimating the direct and indirect water use of tourism in the eastern Mediterranean. J Environ Manage 114:548–556. https://doi.org/10.1016/j.jenvman.2012.11.002
Hardin G (1968) The tragedy of the commons. Science 162:1243–1248
Healy R (1994) The "Common Pool" problem in tourism landscapes. Ann Tour Res 21(3):596–611
Herrera RS (2014) How much longer will the country's water last? Envion 4895. http://www.envio.org.ni/articulo/4895
Hof A, Schmitt T (2011) Urban and tourist land use patterns and water consumption: evidence from Mallorca, Balearic Islands. Land Use Policy 28:792–804. https://doi.org/10.1016/j.landusepol.2011.01.007
IPCC (2007) Summary for Policymakers. In: Solomon S et al (eds) Climate Change 2007: The physical science basis: contribution of Working Group I to the Fourth Assessment Report of the intergovernmental panel on climate change. Cambridge University Press, Cambridge, UK, pp 1–18
Klinger M (2014) The Coveted Coast. Surfer 55(3):66–75, 108
Kuzdas C, Warner BP, Wiek A, Vignola R, Yglesias M, Childers DL (2016) Sustainability assessment of water governance alternatives: the case of Guanacaste Costa Rica. Sustain Sci 11:231–247. https://doi.org/10.1007/s11625-015-0324-6
LaVanchy GT (2017) When wells run dry: water and tourism in Nicaragua. Ann Tour Res 64:37–50. https://doi.org/10.1016/j.annals.2017.02.006
LaVanchy GT, Taylor MJ (2015) Tourism as tragedy? Common problems with water in post-revolutionary Nicaragua. Int J Water Resour Dev 31(4):765–779. https://doi.org/10.1080/07900627.2014.985819
LaVanchy GT, Romano ST, Taylor MJ (2017) Challenges to water security along the "Emerald Coast": a political ecology of local water governance in Nicaragua. Water 9:655. https://doi.org/10.3390/w9090655
Llano-Arias V (2015) Community knowledge sharing and co-production of water services: two cases of community aqueduct association in Columbia. Water Alternat 8(2):77–98
Mehta L, Veldwisch G, Franco J (2012) Introduction to the special issue: water grabbing? Focus on the (re)appropriation of finite water resources. Water Alternat 5(2):193–207. https://doi.org/10.1016/j.landusepol.2006.05.009
OECD (2012) Water governance in Latin America and the Caribbean: a multi-level approach. OECD Publishing, OECD Studies on Water

Ostrom E (1990) Governing the commons: the evolution of institutions of collective action. Cambridge University Press, Cambridge

Pons D, Taylor MJ, Griffin D, Castellanos E, Anchukaitis KJ (2017) On the production of climate information in the high mountain forests of Guatemala. Ann Assoc Am Geogr 107(2):323–335. https://doi.org/10.1080/24694452.2016.1235481

QUENCA (2011) Estimación de la Disponibilidad de Agua Subterranea para Suminstro del Proyecto Guacalito de las Isla. Abril 2011. QUENCA Consulting Group S.A. Managua, Nicaragua

Republic of Nicaragua (2007). Law 620: National Water Law and Enabling Regulations. *La Gaceta*, Number 169, September 2007

Rico-Amoros AM, Olcina-Cantos J, Sauri D (2009) Tourist land use patterns and water demand: evidence from the Western Mediterranean. Land Use Policy 26:493–501. https://doi.org/10.1016/j.landusepol.2008.07.002

Romano ST (2016) Democratizing discourses: conceptions of ownership, autonomy, and 'the State' in Nicaragua's rural water governance. Water Int 41(1):74–90. https://doi.org/10.1080/02508060.2016.110770

Romano ST (2019) Transforming Rural Water Governance: The Road from Resource Management to Political Activism in Nicaragua. University of Arizona Press, Tucson, AZ, USA

Schouten T, Moriarity P (2008) Community water, community management: from system to service in rural areas. Practical Action Publishing, Warwickshire, UK

Solanes M, Gonzalez-Villarreal F (1999) The Dublin principles for water as reflected in a comparative assessment of institutional and legal arrangements for integrated water resources management. Global Water Partnership (GWP), Technical Advisory Committee (TAC), Stockholm, Sweden

Van Noorloos F (2011) Residential tourism causing land privatization and alienation: new pressures on Costa Rica's coasts. Development 54(1):85–90. https://doi.org/10.1057/dev.2010.90

Webster TC, Markley B, Waite L (2001) Water resources assessment of Nicaragua, US Army Corps of Engineers. http://www.bio-nica.info/biblioteca/Webster2001NicWater.pdf

World Bank. (1998) Integrated water resources management in Latin America and the Caribbean. Technical Study No ENV–123, Washington, DC, December 1998

Chapter 4
Imagining the "Emerald Coast": Insecure Land Tenure, Property Disputes, and Tourism Development

Abstract Since the mid-1990s, Nicaragua's coastal areas have been increasingly transformed into places of tourism and leisure. This transformation often displaces rural populations who have imbued these spaces with their own social, political, and historical meanings. Natural resources and coastal land are now valued differently and targeted for different purposes than in the past. For example, land re-distributed to agricultural cooperatives by the Sandinista government is now seen as valuable spaces of tourism and leisure by national and international elites, resulting in conflict with regard to access and control over resources central to the livelihoods of local families. This chapter examines how processes of tourism and real estate development are transforming land tenure along the southwestern Pacific coast of Nicaragua. Using qualitative research, we analyze how these processes are changing access to and control over land and natural resources along Nicaragua's coastal areas. We argue that the immense complexities surrounding land tenure in Nicaragua are exacerbated by continued tourism and real estate development in coastal areas. We look at the cases of two specific tourism-related land conflicts in order to better understand how these conflicts emerge and develop on the "Emerald Coast."

Keywords Nicaragua · Tourism · Land tenure · Land use change · Elites

4.1 Introduction

Since the mid-1990s, Nicaragua's coastal areas have increasingly transformed into places of tourism and leisure. This transformation often displaces rural populations from the places that they imbue with their own social, political, and historical meanings. Natural resources and coastal land are now valued differently and targeted for different purposes than in the past. For example, land redistributed to agricultural cooperatives by the Sandinista government is now seen as valuable tourism and leisure space. This has caused conflicts with regard to access and control over resources central to the livelihoods of local families, as well as conflicts of property rights to illustrate one of the broader process of land grabbing taking place in the region (Aguilar-Støen 2016).

G. T. LaVanchy et al., *Tourism in Post-revolutionary Nicaragua*,
SpringerBriefs in Latin American Studies,
https://doi.org/10.1007/978-3-030-55632-7_4

In this chapter we examine how processes of tourism and property development are transforming land tenure along the southwestern Pacific coast of Nicaragua. We analyze how these processes change access to and control over land and natural resources in Nicaragua's coastal areas. We argue that the immense complexities surrounding land tenure in Nicaragua are exacerbated by continued tourism and real estate development in coastal areas. We look at the cases of two specific tourism-related land conflicts in order to better understand how these conflicts emerge and develop on the "Emerald Coast." Our analysis focuses on local fishermen and their families, actors involved in tourism and real estate development, and the Nicaraguan State. With our analysis we wish to better understand ways in which coastal landscapes and livelihoods are being re-shaped by tourism development and the changing political economy of Nicaragua.

4.2 The Historical Land Context in the Area

The current configuration of private land tenure in Gigante has its origins in three processes: (1) the opening of Nicaragua to tourism, (2) the transition of informal titles to regularized individual titles that could be sold on the open market, and (3) the illegal dissolution of the Pedro Joaquin Chamorro agricultural cooperative by tourism developers with the express purpose of gaining control of land held communally by the cooperative. Land with prime coastal views and beach access was often viewed by the cooperatives as land with little value until the arrival of new values and money involved in the tourism boom along the coast. Real estate developers and tourism investors were interested in land "with a view," which placed new values on land considered less desirable and productive by local people (Espinosa 2007).

The area now known as Gigante was part of one of Nicaraguan dictator Anastacio Somoza's private estates, called Finca Güiscoyol, covering 4500 ha of land. Somoza acquired the land in 1937, which was used for cattle ranching until its confiscation in 1979 during the Sandinista land reform (USAID 2011). Through the land reform, close to 840 ha of the finca Güiscoyol were re-distributed to a group of landless peasants from Tola and their families in the Pedro Joaquin Chamorro Agricultural Cooperative. The cooperative received its official title in 1994 and each family was awarded individual usufruct rights to the land through a "posesión" within the land collectively owned by the cooperative. Policy changes brought by the Violeta Chamorro administration (1990–1997), which were focused on privatization and individualization of property rights, then opened up the possibility to sell the cooperative land (Everingham 2008).

This re-distribution of land to an agricultural cooperative was unique in the immediate Gigante area. Many of Nicaragua's wealthiest families supported the removal of Somoza from power and were initially spared the confiscation of their assets (Bull et al. 2014). In the Pacific region not all land was confiscated, and most land remained in large privately owned export estates (Brockett 1998). Indeed, some of the land in Tola that eventually became tourism resort developments never left private hands.

The Chamorro government also restored the rights of some landowners whose land had been confiscated by the Sandinista government. At the same time, poor families were promised that they could keep the land they had so newly acquired through the Sandinista land reform. To further complicate land-holding issues, at the end of the Contra War former FSLN and CONTRA soldiers were also promised land of their own. This new cycle of land distribution often involved land that was already allocated to—and in some cases even titled in the name of—beneficiaries of the Sandinista land reform. These shifting policies gave rise to competing land claims, tenure insecurity, and conflicts. Many of these conflicts became violent (Broegaard 2009; Dillie 2012). Land ownership again became highly concentrated in areas where large landowners successfully asserted their prior rights and dispossessed beneficiaries of the Sandinista reforms (USAID 2011). Both the Sandinista state-led land reform, and the post-Sandinista counter-reform were highly politicized and produced neither the expected nor the desired re-distribution of land, and land ownership in Nicaragua continues to be highly concentrated (Broegaard 2009).

4.3 Land Conflicts and Tourism Development

Even before cooperative members received their individual land titles, land purchases related to tourism development had already begun in the area. In the late 1990s, the Pedro Joaquin Chamorro Cooperative of Gigante began to seek individual titles for their plots of land within the cooperative so that they could potentially sell the land. However, tourism development came to a region with a long history of extraordinary concentration of land ownership, and with complex and contradictory history of land ownership. Tourism, and the associated re-concentration of land in the hands of tourism operations, exacerbates existing patterns of unequal and insecure land tenure, especially as land in the area becomes increasingly valuable. The situation is then further exacerbated by the increased juridification of social relations[1] that have facilitated the change of land from the hands of locals to outsiders. The increased juridification of social relations does not mean that the legal system is better equipped to solve conflicts in a just manner. Rather, more powerful actors are in an advantageous position to be victorious in legal processes. The outcomes of judiciary processes are often highly politicized and Nicaraguan courts are in practice partisan instruments at the service of Sandinista or Liberal interests (Martínez Barahona 2012).

Land conflicts in the region often involve tourism and real estate developers, former military members (both Sandinista and Contra), and former cooperative members. The conflicts have involved contradictory property claims, power abuse and corruption, falsified property titles, and squatting. Some of the conflicts have been very high profile, some are ongoing, while others have been settled, although a lingering sense of injustice and betrayal still remains among those who lost. In

[1] That is, the increased degrees to which social relations are being textured by formal legal rules. (Couso et al. 2010).

addition to high-profile land disputes there are multiple smaller conflicts involving locals, some of whom are landless and have turned to squatting. Below we present two land conflicts from the Gigante area. Both of these conflicts showcase the immense complexities surrounding land tenure in Nicaragua, which are amplified with continued tourism and real estate development in the area.

4.3.1 Playa Amarillo Conflict

The Playa Amarillo conflict involves land that was part of Somoza's Güiscoyol estate. The land at the heart of the conflict lies just north of Gigante, bordering the Pacific Ocean to the west, and stretches east to the Salinas-Tola highway and the community of El Tambo. It borders Playa Colorado and the Hacienda Iguana residential community to the north. The land includes a 3 km stretch of beach and land behind the beach. The conflict involves private investors, former military members of the National Sandinista Army and the Nicaraguan Resistance, local residents, and the Nicaraguan state through the Attorney General's office.

In 1993, CORNAP[2] reportedly sold or transferred the land to a group of former soldiers, who then sold the land to a private investor, Bayardo Argüello, who intended to develop the land for tourism and real estate development projects. However, in the following years many different titleholders emerged and in 2006 there were at least 18 different property titles for this same piece of land (El Nuevo Diario 2012). One party claiming the land is a ten-person group of former Sandinista soldiers who claim to have received the land in 1988 through the Sandinista agrarian reform. However, the group did not receive a legal property title until 2005. The receipt of this title led to an extensive legal battle that went back and forth through the courts, but eventually ended up with the title being annulled. The group, seen in Fig. 4.1, however, continues to proclaim their right to the land and lives on a section of the land. Group vision and cohesion is largely the result of the leadership of Thomas Borge, also known as "El Bigote" or "The Moustache." He is very welcoming to outsiders and generously relates his involvement in the Sandinista Revolution, the Sandinista fight against the CONTRAS, and the occupation of land by ex-combatants (Fig. 4.2).

Interviewees living on this contested land saw two potential outcomes in the Amarillo conflict; either the government would give them what they see as their rightful property title, or the government pays them compensation for loss of the land. In the case of the Gigante locals, who also lay claim to the land, there was no doubt that this land is rightfully theirs as they see themselves as the rightful beneficiaries of the Sandinista land reform. However, instead of wanting the land for agricultural purposes, it was also very clear that most of the people we spoke with wanted to be able to sell the land to tourism investors once they received a legal property title. This exemplifies how land that was formerly of little interest

[2]Corporación Nacional del Sector Público: the state holding company responsible for managing state property and sales in the 1990s.

Fig. 4.1 The group of ex-combatants who occupy part of the Playa Amarillo area. Here they are seen celebrating the 30th anniversary of "El Triunfo" of the revolution on July 19, 2009. *Source* M. J. Taylor

becomes sought-after and contested because of its increased value through tourism development. The Amarillo case also showcases how the contradictory land policies of the 1980s and the 1990s have resulted in competing land claims, tenure insecurity, and conflicts.

4.3.2 Caso Tola Conflict

The conflict between members of the Pedro Joaquin Chamorro agricultural cooperative and several investors dates back to the year 2000 when the investment group González-Bolaños[3] with ties to former president Enrique Bolaños acquired land from the cooperative. The land in question is now home to the Buccaneer Point residential community and the Aqua Wellness Resort.

In the late 1990s investors became interested in cooperative land for tourism and real estate-related purposes. At this time the cooperative did not have the financial

[3]La Sociedad Inversiones González Bolaños, S.A. including members Alejandro Bolaños Davis, his wife, his two sons, and his son-in-law, Armel González Muhs.

Fig. 4.2 El Bigote of Playa Gigante, Nicaragua. *Source* Madeline Meredith

means to begin the process of obtaining individual titles. The investors and the cooperative struck a deal in which the investors would help with the legal proceedings of surveying and titling the cooperative land. According to interviews with cooperative members, the investment company offered to pay for titling the land in exchange for two landlocked plots within the cooperative.

In the 1990s there were restrictions on the sale of the land because of it being cooperative land. To get around these restrictions, family members of Armel González and Alejandro Bolaños Davis were made members of the cooperative, despite not being landless peasants and beneficiaries of the agrarian land reform. This illegal action allowed them to begin the process of dissolving the cooperative to enable their land acquisition (Fig. 4.3).

However, unbeknownst to the cooperative members, the developers titled lucrative ocean view plots to themselves instead of the landlocked plots they had previously agreed upon. According to testimony in the Nicaraguan Supreme Court and based upon our interviews, the cooperative sold the land to the investment group in 2000. Sometime between 2000 and 2003 the land acquired by the González-Bolaños group was sold and transferred to a different company, which included foreign tourism

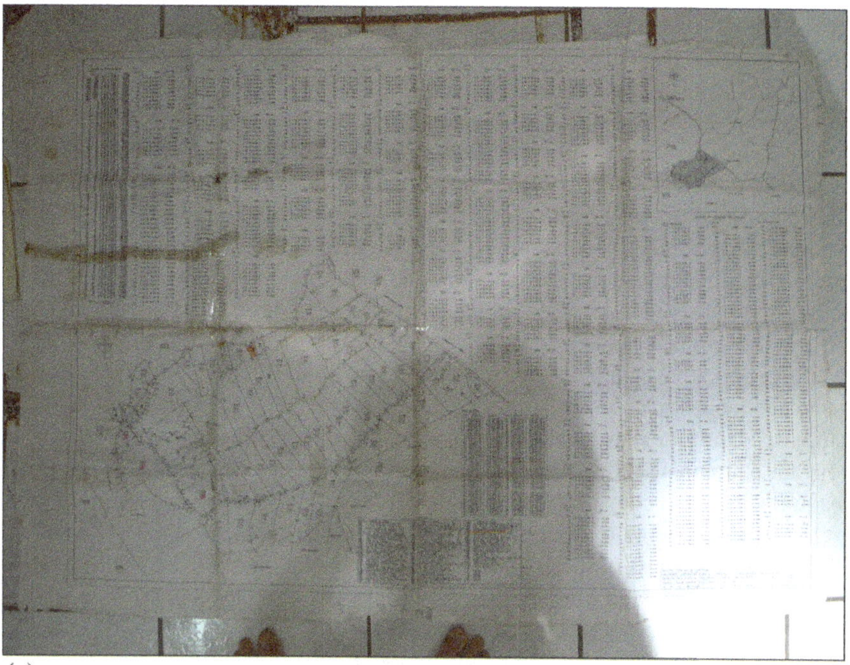

(a)

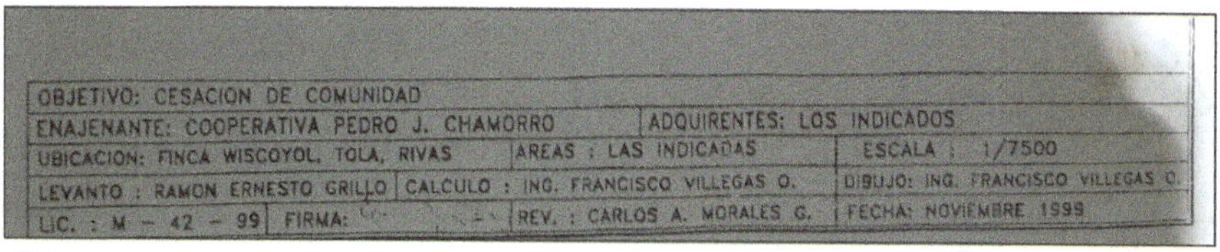

(b)

Fig. 4.3 **a** Plans of the Pedro Joaquín Chamorro agricultural cooperative showing individual plots surveyed as part of the land grabbing scheme by the investment group González-Bolaños. **b** Its title, "cessation of the community" can be seen in the zoom view of the plans. This cessation of the community was an illegal move by the land developers and their lawyers, done without permission of the cooperative. *Source* M. J. Taylor

and real estate developers.[4] This became the group of investors who would begin the development of Arenas Bay (later called Aqua Wellness Resort and Buccaneer Point). The process resulted in a fierce conflict and legal battles that are still not resolved.

In 2004, a group of members from the original cooperative filed a lawsuit against the investment group to annul the land sale and return the land into their hands.

[4]*Sociedad San Cristóbal, S.A.* with shareholders; husband and wife Jocelyn Carnegie and Gail Geeriling, Alejandro Bolaños Davis, his daughter Alejandra Bolaños Chamorro, and Armel González Muhs.

They claimed that the official contract of sale for the land had been vague and imprecise, and had failed to mention which plots of land were being sold or their actual boundaries.[5] The legal battle reached its peak in 2006 and 2007 with highly politicized corruption allegations. Following the allegations the government appointed a special commission to delve into the case (Pantoja 2007). The commission concluded that there was sufficient evidence to support claims of unlawful acts committed by the González-Bolaños group and several public officials, with regards to their dealings with the cooperative. Unfortunately, this did not help the members of the cooperative in their legal battle against the investors and today they have all but given up, not having the means to "play ball" in an expensive legal system, where the investors have a clear home advantage.

The cooperative members have made it clear that they do not wish to evict current home and landowners on the contested land. They are in agreement that those who purchased land from the developers did so not knowing about the conflict. What they do want is compensation from the developers. As we see in the case of the Pedro Joaquin Chamorro cooperative versus the Arenas Bay developers, it is often those with sufficient economic or political resources who can claim, formalize, and enforce land rights, regardless of the original legal tenure situation of the land.

To better understand how these conflicts emerge and develop, we will discuss the factors that have contributed to the production of the Emerald Coast, both as an imagined place and as a material reality, in the following section.

4.4 Imagining the Emerald Coast

Tourism in and around Gigante evolved rapidly over the past decade. Californian surfers-turned entrepreneurs started some of the first tourism businesses in Gigante in 2004 and 2005. Many of them wanted to re-create a "Californian" way of living (their own words). In doing so, they imagined Tola as a frontier, where those who have the means can create a place that suits their lifestyle. These tourism pioneers contributed to infuse the coast with new meaning—a meaning quite different to that previously held by the Nicaraguan locals.

Words like untouched, undiscovered, and unspoiled are commonplace in descriptions of Tola's "Emerald Coast," as if Nicaragua was a no man's land before the arrival of tourism. Nicaragua is being re-invented as a new frontier for tourists and investors of all shapes and sizes, from the surfers and the backpackers to the expats and retirees looking to settle, to the high-end tourist looking for an exclusive and lavish experience.

The image of the Emerald Coast is advertised and marketed mostly toward North Americans and Canadians interested in living, retiring, and investing overseas

[5]This is according to a report by a special commission appointed by the Nicaraguan National Assembly in order to investigate the case (Pallaís Arana 2007). The report found numerous irregularities regarding the land sale.

Fig. 4.4 Branding and advertising of the Emerald Coast by tourism developers. *Source* G. T. LaVanchy

(Fig. 4.4). This marketing depicts the Emerald Coast as being an affordable "piece of paradise" and a "more affordable" Costa Rica (Levius 2017). It has become a place where tourists and expats can "make their dreams come true" and "maybe make a lot of money" (Lloyd 2006).

National and international companies have been involved in the development of resorts and residential communities in Tola, such as Iguana Golf and Beach Club, Arenas Bay Development, Guacalito de la Isla, and Rancho Santana. To illustrate, International Living bought over 1000 ha of land in Tola, which then became the Rancho Santana residential community, a high-end resort and residential community north of Gigante. The New York Times has called Nicaragua "the next luxury destination" and CNN referred to the country as the "new eco-travel darling of Central America." Aqua Wellness Resort has been certified by the Rainforest Alliance, and they present the project's vision as: "Aqua's explicit vision statement is to lead luxury sustainable development in Nicaragua….we have conceived, built, and operated our resort in close partnership with our (sic) local communities in Gigante and Tola and in concert with nature." This statement of course conceals the fact that the ownership of the land where the resort has been built is disputed and that the land was acquired through arguably questionable means. Instead, it presents an image of eco-friendliness and social responsibility.

In addition to the surfers and entrepreneurs, the Government of Nicaragua, through INTUR (Nicaraguan Tourism Board) and in alliance with Carlos Pellas, the owner of Guacalito de la Isla, is promoting Tola and the Emerald Coast as a high-end "ecological" tourism destination. According to Guacalito's web site "The spirit of Guacalito de la Isla aligns with the elegant call of the ocean waves, creating an environmentally conscious, luxury retreat that merges twenty-first century comforts with Central America's unique cultural mosaic." Guacalito de la Isla is often presented as a project of "ecological sustainability" that will benefit the whole country. The opening of the first phase of Guacalito de la Isla, coincided with a publication in The New York Times (Felsenthal 2012) where Nicaragua was ranked the third most exciting tourist destination in the world, out of a list of 46 places. Guacalito de la Isla and the Emerald Coast have been featured in numerous publications including the New York Times, CNN, The Wall Street Journal, Forbes, Vogue and others.

Not only have these developments physically altered the landscape of Tola and Gigante through the construction of resorts, residences, and infrastructure. They have also influenced and shaped the way Tola and Gigante are imagined. In this eco-luxury narrative, the lives and livelihoods of the locals are rendered invisible and the coast is presented as a paradise to be conquered by those with money and eco-friendly aspirations.

4.4.1 Legislative and Institutional Factors

The tourism sector in Nicaragua began to expand rapidly in the 2000s as it did in other parts of Central America (Babb 2010; Cañada 2010; Hunt 2011; Hunt and Stronza 2011). While the coast appeals to the desires of retirees and expatriates to move to far-away corners of the world in order to buy a piece of paradise or live out eco-luxury dreams, it also represents an investment opportunity. Important triggers for these developments have been increased global connectivity and policies aimed at attracting foreign direct investment.

Government commitment in Nicaragua to developing tourism is reflected in a cluster of tourism laws that have encouraged sectorial growth with tax holidays and tariff concessions. Including in these laws are the Law on Incentives for the Tourism Industry (Law 306 of 1999), the Law of Resident Pensioners and Retirees (Law 694 of 2009), and the Law for the Development of Coastal Areas and its Regulations (Law 690 of 2009). Law 306, passed in 1999, declares tourism as an "industry of national interest," and is considered by investors and developers as one of the most generous of its kind in Central America, offering broad tax exemptions. Law 694 serves similar purposes as the Law 306 and was intended to promote residential tourism through various tax exonerations and exemptions (INTUR 1999, 2009).

Law 690 on the development of coastal areas filed in 2005 was passed in 2009 following a long discussion process about the delimitation of public and private areas along beaches and the manner of access to them, as well as the zoning criterion to be used (INTUR 2009). In the end, the law ended up being mostly beneficial to tourism-residential interests and provides judicial security for private properties on the coast. The law established that public use reaches 50 m inland from the mean high tide line. This means de facto privatization of the public domain that the government had over a two kilometer-wide band along coastal areas, reducing it to 50 m. While the law states that its objectives are to "guarantee access by the population to the coastline…in order to guarantee use and enjoyment of the coast by the population," (INTUR 2009) this has, in general, not been the case. The continued proliferation of resort and residential tourism developments along the coast restricts the maneuvering space of many local people living in coastal areas.[6]

Tourism incentive policies have played a large role in the tourism and real estate development along Tola's coastline and the privatization of land, restricting access to and control over resources, and dispossessing local communities. The legal development outlined above reflects that the interests of investors, foreign and local, have gained preeminence over the interests of other groups in the negotiations.

4.4.2 *Elite Dynamics*

Elite dynamics are intertwined in the development of the Emerald Coast in three ways. First, elites act as financial actors securing capital for the development of tourism projects. Second, elites promote and influence changes in legislation and policies that favor investment. Third, by close association with the political elite, business elites in Nicaragua have been able to use decentralized deal making to secure exemptions and loopholes that favor their interests.

Alliances between the economic elite and the Sandinista political elite are nothing new. The Sandinista government of the 1980s used a strategy of broad political alliances, including the elite, to govern after the revolution (Vilas 2001). Many members of the Nicaraguan economic elite participated in the FSLN and after the triumph of the revolution they came to occupy high ranks in the Sandinista government. Ruptures and continuities mark the relationship between the economic elite and the Sandinistas. Ruptures in the past were, among other things, colored by disagreements in regard to the land reform. Processes of economic liberalization and globalization have fostered the development of a new economic elite in Nicaragua and the re-composition of the old elite (Spalding 2013). The emerging alliances and interfaces between the FSLN and the government of Daniel Ortega on the one hand, and the economic elite on the other, have to be understood within a context of shifting international alliances and policy adjustments, as well as part of Ortega's tactics to secure control over the entire state apparatus (Martí i Puig 2016).

Within this context, and with continued growth of tourism in the Central American region, tourism development is discursively constructed as a splendid economic option for the country, and different initiatives are launched to allow Nicaragua to participate in a globalized tourism boom. The involvement of the economic elite in shaping tourism policy and legislation represent one of the continuities. In 2013, President Ortega sent a proposal for changes to the Constitution. Among those changes was one that proposed that economic decisions in Nicaragua be taken in consensus with the economic elite, through so-called sectorial cabinets. The Superior Council of

[6]The coastal law No. 690 clearly states that developments must provide access routes to the coastline through easements. An easement is an access route to the coastline that coastal plots of land are compelled to open in order to guarantee use and enjoyment of the coast by the population. Investors must include the easement in the initial design of their tourist project (INTUR 2009, 4).

Private Enterprises (COSEP) would be part of these cabinets with the responsibility to govern economic issues in the country. In the past, the participation of the economic elite occurred through their involvement in high-ranking offices of the government, which necessitated their direct involvement with political parties. Now these alliances are institutionalized in the sectorial cabinets. A strategy of "consensus" by way of which economic elites have attained ample influence in economic matters, and at the same time, secured political stability in the country is now dominant.

As Bull et al. (2014, 174) observe, for business groups in Nicaragua, the period of Daniel Ortega's presidency has been characterized by a "gentlemen's agreement." Ortega has been able to tighten his grip on the Nicaraguan state without opposition from the business groups and has in exchange implemented pro-business policies and consulted with them before proposing any important legislation. Indeed, the Superior Council of Private Enterprises (COSEP) has become the main forum for economic policy in the country, making the elite able to influence most of Ortega's economic policies. Carlos Pellas, owner of Grupo Pellas and Guacalito de la Isla, often emphasizes Ortega's ability to steer the economy and maintain a good dialog with the business sector and labor unions (Bull et al. 2014).

In Gigante and Tola, some of Nicaragua's powerful elite families, or foreigners with connections to them, own an increasing number of tourism development projects. Our empirical material indicates close links between the Nicaraguan business elite and political elite suggesting that they cooperate to pursue the development of luxury tourism in the area. Central figures are Carlos Pellas, member of Nicaragua's most prominent elite families, Bayardo Arce, Ortega's most influential economic advisor, and his relatives or friends, and former war hero and member of the FSLN Eden Pastora. Arce is commonly identified with the pro-business wing of the FSLN and who is currently the economic advisor of President Daniel Ortega (Spalding 2013). Arce is known for being very pragmatic in his approach to business elites and economic growth (Rogers 2011).

Guacalito de la Isla is one of the larger tourism and real estate development projects surrounding Gigante and opened its doors in early 2013. The 500 million dollar mega-development covers 650 ha of land and is self-proclaimed as the most exclusive development of its kind in Nicaragua, if not Central America. The project was developed by Pellas Development Group, which is a subsidiary of Grupo Pellas, one of the largest financial conglomerates in Central America, and controlled by arguably the most powerful family in Nicaragua, the Pellas family.[7] The Pellas family has a long and established presence among Nicaragua's business elite (Bogan 2009). Not only are the Pellas family in an exceptional position to develop tourism and real estate along Tola because of their enormous wealth, they are also in a position to exert

[7] Grupo Pellas has business interests, among other things, in: sugar, ethanol, rum, banking, energy, car dealership, media, telecommunications, health care, and recently, high-end tourism. Pellas Development Group was formed in 2005 as a specialized unit that focuses on real estate investments and development across Central America. Apart from Guacalito de la Isla in Nicaragua, Pellas Development Group have two other projects in the region; the 1500 ha Santa Elena Preserve in Guanacaste, Costa Rica; and the 283 ha Santa María Golf and Country Club in Panama (Pellas Development Group 2016).

power over the market and influence tourism legislation and the actual investment setting. Grupo Pellas had a large role in the building of the recently opened (as of October 2015) airport in Tola, the "Emerald Coast Airport," with generous tax exemptions from the government (El Nuevo Diario 2014; La Voz del Sandinismo 2014).

Elites, through their control over resources and often the institutions that regulate their use, are in a unique position to influence the way in which land use is re-shaped, and how resources are accessed, used, and controlled. Paying attention to elite dynamics is important for understanding how tourism is transforming Gigante and Tola's coastal areas, and the ways in which uneven access to certain resources allows for different actors to participate in and benefit from tourism in the area.

4.5 Conclusion

Gigante and Tola's coastal areas have increasingly transformed from being places imbued with social, political, and historical meaning for local people to places of leisure for tourists, expatriates, and the rich. Not only are Gigante and Tola being reproduced as a tourism space, they are also transformed into part of the "Emerald Coast," which is increasingly a space that excludes the original inhabitants of the region. This re-imagination of Tola's coastal areas as a tourism space has re-shaped land tenure in the area and is causing conflicts with regard to access and control over resources, which are central to the livelihoods of local families. The tourism and real estate boom has in many cases amplified the already precarious situation of land tenure insecurity in the area. We argue that the ability to negotiate conflicting interests is, to a large extent, constrained by the resources available to different actors to shape and influence the production of space.

Access to land, the coastline and the ocean is vital for the local population in Gigante and Tola's coastal areas. By restricting access to parts of the coastline through the privatization of land and the creation of gated resort and residential communities, tourism directly affects local people's livelihoods, both with respect to their income and their subsistence. Real estate development transforms land tenure and access to and control over resources. This becomes particularly relevant when considered in the context of social inequality and the unequal distribution of resources and land throughout rural areas in many parts of the world.

The perspective of local inequality through transformed land tenure is perhaps best understood through the local voice. Following are words of a member of the Pedro Joaquin Chamorro Cooperative, Juan Ramon. We sat with him, along with fellow cooperative member Rufino Lanza Zapata, one June night to share stories and local food. Juan Ramon pulled out his guitar and gave us his perspective on the development that has taken place in the Gigante area (Fig. 4.5).

Fig. 4.5 Juan Ramon from Gigante sings “Mi Papa Tenía un Terreno” while Rufino Lanza Zapata (left) reflects and listens. June 2012. *Source* M. J. Taylor

Mi papá tenía un terreno
Y un rico se lo quitó
Lo mandaron a la calle
Aguantando sol y agua
Y llorando me quedé yo
Solo Dios puede mirar
En las tierras que nos (los) dan
Pa’ qué tantas escrituras?
Pa’ qué tanto brujuleo?
Papelismo y nada más

Cuánto gana un coronel?
Cuánto gana un licenciado?
Cuánto gana un diputado?
Compra carros bien bonitos
Con aire acondicionado
Y le sobra para comer
Y el pobre campesino
Aguantando sol y agua
Pa’ poderse mantener

Solo Dios puede mirar
En las tierras que nos (los) dan
Pa' qué tantas escrituras?
Pa' qué tanto brujuleo?
Papelismo y nada más

English translation

My dad had some land
And a rich person took it
And left him in the street
Under the sun and the rain
And left me crying
Only God can tell
what will happen with this land
Why so many land titles?
Why so much trickery?
Bureaucracy, nothing more!

How much does a general earn?
And a person with degrees?
And a senator?
They buy very nice cars
With air conditioning
And they still have money to eat
And us, the poor peasants?
We are left under the sun and rain
to struggle on

Only God can tell
what will happen with this land
why so many land titles?
Why so much trickery?
Bureaucracy, nothing more!

Limited access to social, economic, and political resources may prevent the local population from obtaining necessary support from public institutions and thus benefit from tourism. A weak and very expensive legal system often precludes effective and equal enforcement of legislation. Foreign and domestic elites, who generally have greater access to these resources, are in a privileged position to not only outcompete locals in tourism but are also in a position to influence how locals are able to, or are excluded from, participating in tourism. The production of the Emerald Coast as a tourism space illustrates the complex web of actors and social relations occurring at multiple scales, which are leading to an increasing "elitization" of land, and which is in many cases reproducing inequalities between local people, communities, regions, and nations.

References

Aguilar-Støen M (2016) Beyond transnational corporations, food and biofuels: the role of extractivism and agribusiness in land grabbing in Central America. Forum for Development Studies 43(1):155–175. https://doi.org/10.1080/08039410.2015.1134641

Babb FE (2010) The tourism encounter: fashioning Latin American nations and histories. Stanford University Press, Stanford, Calif

Bogan J (2009) Navigating Nicaragua. *Forbes*, October 20, 2009. http://www.forbes.com/2009/10/20/grupo-pellas-nicaragua-business-chamorro.html

Brockett CD (1998) Land, power, and poverty: agrarian transformation and political conflict in Central America, 2nd edn. Westview Press, Boulder, Colo

Broegaard R (2009) Land access and titling in Nicaragua. Development and Change 40(1):149–169. https://doi.org/10.1111/j.1467-7660.2009.01509.x

Bull B, Castellacci F, Kasahara Y (2014) Business groups and transnational capitalism in Central America: economic and political strategies. Palgrave Macmillan, New York, NY

Cañada, E. (2010). Tourism in Central America, social conflict in a new setting (No. 32). Alba Sud. Retrieved from http://www.albasud.org/publ/docs/32.en.pdf

Couso J, Huneeus A, Sieder R (eds) (2010) Cultures of legality: judicialization and political activism in Latin America. Cambridge University Press, Cambridge; New York

Dillie B (2012) Ill fares the land: The legal consequences of land confiscations by the Sandinista government of Nicaragua 1979–1990. PhD Dissertation, The London School of Economics and Political Science, UK. http://core.ac.uk/download/pdf/9579886.pdf

El Nuevo Diario (2012) Las codiciadas costas del sur de Tola. El Nuevo Diario, February 21, 2012. http://www.elnuevodiario.com.ni/nacionales/242502-codiciadas-costas-sur-tola/

El Nuevo Diario (2014) Tola tendrá aeropuerto. El Nuevo Diario, February 06, 2014. https://www.elnuevodiario.com.ni/nacionales/310261-tola-tendra-aeropuerto/

Espinosa F (2007) Consultoría para la elaboración de un estudio catastral en el litoral marino de Tola. MASENACE-GTZ

Everingham M (2008) Agricultural property rights and political change in Nicaragua. Latin American Politics and Society 43(3):61–94. https://doi.org/10.1111/j.1548-2456.2001.tb00179.x

Felsenthal J (2012) A delicate balance. Nicaragua. T Magazine. The New York Times Style Magazine, November 18, 2012. http://tmagazine.blogs.nytimes.com/2012/11/16/a-delicate-balance/?_r=0

Hunt C (2011) Passport to development? Local perceptions of the outcomes of post-socialist tourism policy and growth in Nicaragua. Tour Plan Dev 8(3):265–279. https://doi.org/10.1080/21568316.2011.591155

Hunt C, Stronza A (2011) Missing the forest for the trees?: incongruous local perspectives on ecotourism in nicaragua converge on ethical issues. Hum Organ 70(4):376–386. https://doi.org/10.17730/humo.70.4.xj187458416w1gr8

INTUR (1999) Law No. 306. Law on Incentives for the tourism industry of the Republic of Nicaragua (Law No. Official bulletin No. 117). Instituto Nicaragüense de Turismo (INTUR), Nicaragua

INTUR (2009) Law No. 690. Law for the development of coastal áreas and its regulations (Law No. Official bulletin No. 141). Instituto Nicaragüense de Turismo (INTUR), Nicaragua

La Voz del Sandinismo (2014) Construirán aeropuerto internacional en Tola, February 06, 2014. https://www.lavozdelsandinismo.com/nicaragua/2014-02-06/construiran-aeropuerto-internacional-en-tola/

Levius T (2017) The Emerald Coast: Nicaragua's hidden jewel. CNN, September 15, 2017. http://edition.cnn.com/travel/article/nicaragua-emerald-coast/index.html

Lloyd C (2006) International living has created an empire selling customers paradise. SFGate, January 13, 2006. http://www.sfgate.com/entertainment/article/International-Living-has-created-an-empire-2544013.php

Martí i Puig S (2016) Nicaragua: Desdemocratización y caudillismo. Revista de Ciencia Política (Santiago) 36(1):239–258. https://doi.org/10.4067/S0718-090X2016000100011

Martínez Barahona E (2012) A politicized judiciary. In: Close D, Martí i Puig S, McConnell SA (eds) The Sandinistas and Nicaragua since 1979. Lynne Rienner Publishers, Boulder, CO

Pallaís Arana JB (2007) Comision de investigacion Caso Tola, Rivas. (No. 20085268). Asamblea Nacional, Managua, Nicaragua. http://legislacion.asamblea.gob.ni/SILEG/Iniciativas.nsf/8e6855fc5b6dce890625733100610704/30e148e06bac11ae062573de00715a89?OpenDocument&ExpandSection=1&TableRow=3.0#3

Pantoja A (2007) Crean comisión que investigará caso Tola. El Nuevo Diario, June 13, 2007. http://archivo.elnuevodiario.com.ni/politica/212462-crean-comision-que-investigara-caso-tola/

Pellas Development Group (2016). http://pellasdevelopment.com

Rogers T (2011) The Nicaragua dispatch. The Nicaragua Dispatch, October 16, 2011. http://nicaraguadispatch.com/2011/10/arce-we-have-to-generate-wealth

Spalding RJ (2013) Business and state relations in post-revolutionary Nicaragua: Elite realignment and the new strategy of collaboration. Flacso-Costa Rica. https://www.american.edu/clals/upload/Spalding_Business_and_State_in_Nicaragua.pdf

USAID (2011) USAID country profile. Property rights and resource governance. Nicaragua. USAID. http://usaidlandtenure.net/sites/default/files/country-profiles/full-reports/USAID_Land_Tenure_Nicaragua_Profile.pdf

Vilas C (2001) Las sociología política latinoamericana y el "caso Chávez": entre la sorpresa y el déjà vu. Revista Venezolana de Economía Y Ciencias Socia 7(2):129–146

Chapter 5
From Boom to Bust and Back Again? Current Dilemmas Related to Tourism Along Nicaragua's Pacific Coast

Abstract This chapter captures the movement of tourism in Gigante from a "boom" to its current "bust" phase. Particular attention is given to the implications of the recent bust on local residents with respect to loss and opportunity related to marine, groundwater, and land resources. Concluding thoughts on the future of tourism are offered.

Keywords Tourism · Political crisis · Economic cycles

5.1 Introduction

Tourism and foreign ownership of tourism operations boomed, despite the hesitation from 2007–2010. Nicaragua became the darling of Central American tourism with the fastest growth rates (although not the highest absolute numbers of visitors) and many outsiders began to call it the "Costa Rica of thirty years ago" or the "new Costa Rica."

In the first decade of 2000, and well up to April of 2018, tourism boomed. Indeed, Nicaragua became a family destination as well as a spring break escape for young North Americans. San Juan de Sur became an international party spot. Nicaragua was posed to become the next "safe but unspoiled" destination with easy access from North America and Europe. This boom in tourism and the concomitant impact of tourism on local lives and environments has been documented in this book. Specifically, we examined the competition over natural resources that accompanied the growth of tourism. We drew attention to the fight for sea space and access to fisheries. We also examined the struggle for water in a poorly understood aquifer and showed how those with more money were able to extract more water out of the ground despite very strong laws that prevent water grabbing—no surprises there! The change in land ownership was documented in this volume and illustrated the power of outside capital to give new meaning to places. Our political ecology approach allowed us to question the differentiated power in social and economic structures and how these differences mapped onto the conflict experiences of stakeholders.

G. T. LaVanchy et al., *Tourism in Post-revolutionary Nicaragua*,
SpringerBriefs in Latin American Studies,
https://doi.org/10.1007/978-3-030-55632-7_5

5.2 Cycles in Tourism

The boom in tourism, largely led by outside investors, provided hundreds of thousands of permanent and temporary jobs for Nicaraguans. This occurred mostly at the lower level, but increasingly locals held middle-level positions and a few upper-level positions. Undeniably, the boom provided benefits to the local economy. In our decade-long research along Nicaragua's Pacific coast, we have seen the tangible outcomes of the economic uptick within Nicaragua, in part due to tourism. Labor laws in Nicaragua are strong, and thus ensure that minimum wage is paid, benefits provided, and termination pay granted. In this manner, the money from tourism did translate into real jobs with real benefits. Local residents, with cash from new jobs, converted old houses made of wood into concrete structures with glass windows and doors. Many families purchased motorcycles, cars, and minibuses. Many Nicaraguans migrated to the coast from other areas of the country, creating a local land market as well as a rental scene.

Life was good. And Nicaraguans themselves reported this as the case—they told us so! Many enjoyed the benefits from tourism and wanted it to continue growing. They reveled in newly paved roads and electricity in their houses. Yes, they expressed concern over increased foreign ownership of land, lamented the domination of foreigners in the restaurant and accommodation realms, but they also maintained active discussions on how they could enter that market—indeed, some did at the lower end for backpackers and local tourists. They also pointed to their drying wells and how large tourist operations appeared to still have plenty. Our research showed otherwise. Basically, all residents along the coast, large operations and small local households, suffered water shortages during the drought years from 2012 to 2018. The drought coincided with the largest growth in tourism infrastructure like golf courses and spas, and thus it was easy for locals to point out the inequalities in water use. Larger tourism operators were, however, able to withstand the drought because of their deeper pockets and ability to drill more wells.

In surveys we conducted in 2014, some locals also expressed concern about the change in culture that tourism brings with it—the mixing of two worlds. Some respondents mentioned the increase in drug and alcohol use, as well as a more sexually permissive party scene. They mentioned this especially as related to surfing tourists. Still others pointed to the positive dimensions of surfing that gave young locals a healthy sport and lifestyle to aspire toward. In general though, most local residents welcomed tourism and tourists. This posture is confirmed by in-depth and repeat interviews we conducted with local residents over the last ten years.

5.2.1 Tourism in Its Current State in Gigante

This boom and its accompanying benefits ended abruptly in April of 2018. Protests by university students over increases in social security taxes and a reduction in benefits paid to pensioners resulted in widespread repression by the government (Martí 2018). In the months after April 2018, almost 400 people were killed by government-sponsored paramilitary groups, thousands were imprisoned and remain in prison, and tens of thousands of Nicaraguans fled into exile. To this day, government repression continues and protest is banned by law. The government continues to arrest and detain citizens involved in any form of protest.

Beyond its immediate impact on citizens, this extreme government oppression and consolidation of power by the Daniel Ortega/Rosario Murillo regime has undermined the tourism industry, which once it promulgated as an economic way forward for the country. Now, North American and European countries have issued firm travel advisories to their citizens not to travel to Nicaragua. Tourism dried up immediately. Empty planes, empty restaurants, empty hotels, empty beaches, and empty waves. A few South Americans, who are perhaps more immune to protest and violence, continued to come on previously booked surf trips. Even that flow, though, soon dried up. Now 80% of small hotels are closed. Now 35% of Nicaragua's restaurants are closed. As of late 2018, over 70,000 people directly employed by the tourism industry have lost their jobs. (This is out of 120,000 that were directly and full-time employed in the tourism industry.) Income from tourism in 2019 is expected to be between 62 and 65% less than in 2017 (Alvarez 2019). The number of people losing jobs continues to increase, although government agencies hide the true numbers in their attempt to persuade visitors that Nicaragua is "back to normal."

"Normal" is far from the reality on the ground. Mukul Resort (inside the Guacalito development led by the Pellas Group), one of the largest employers on the Pacific coast, closed its doors in early June 2018. Smaller operations hung on for a little longer, trying to will the political crisis to end. They too eventually closed their doors because they could no longer afford the hefty tourism taxes paid to the central government.

So, as tourism crashed and employment plummeted, how have locals survived? First, in attempts to keep businesses alive, once exclusive resorts opened their doors in novel ways and encouraged visits to their spaces by many more national and local residents. This allowed locals, for the first time, to access the spaces from which they had previously been excluded. In one sense, this invitation into the space of the resorts is a reclaiming of space that was once theirs, and a space from which they had been excluded because of their ethnicity and economic status. This new arrangement did not last long, however, as the disposable income of locals, often tied to employment within the tourism sector, soon dried up. This newly altered economic status again prevents their visits to exclusive spaces.

Second, small restaurants along the coast, both locally and foreign owned, rotate opening nights to enable partial employment of their employees. Many of the large tourism operations have finally closed. Only one (Rancho Santana) has continued with full employment, using the time to perform deferred maintenance and to build out infrastructure for what they believe will be another boom.

5.2.2 Moving On?

How have locals responded to this new lack of jobs? It is not as though they can migrate to other areas of Nicaragua in search of employment because the whole economy has taken a downturn. Simply then, some Nicaraguans have migrated to Costa Rica. What else? Other coastal residents, who did not sell their land to outsiders, have returned to that land with renewed vigor, planting beans, rice, maize, plantains, and yucca, and raising livestock. Many younger residents of the area have turned to experienced agriculturalists like Rufino Lanza Zapata (Fig. 5.1) for advice on cultivating subsistence crops. Rufino is one of the original members of the agricultural cooperative that received land around Gigante in the 1990s. Such interactions among

Fig. 5.1 Rufino Lanza Zapata. In times of crisis, younger residents have turned to Rufino for advice on cultivating subsistence crops. Rufino is one of the original members of the agricultural cooperative that was granted the land around Gigante in the 1990s. *Source* M. J. Taylor

local residents not only serves to promote food security, it also strengthens social cohesion and local identity.

Unsurprisingly, the real estate market has stalled amidst the decline in tourism and outside interest. This stall has also abated the conflict over land. Data on sales decline are difficult to obtain since many of the real estate offices along the southwest coast have simply closed their doors until more opportune conditions. Many non-Nicaraguans are in a holding pattern and hope to ride out the crisis. They have invested too much in their land, homes, and businesses to sell at a loss. This is not the case for locals, however. Many locals who own lots along the Gigante beach have recently put them up for sale at drastically reduced prices. Lots that had been previously listed in excess of US$100,000 during the era of the "Emerald Coast" boom can now be had for US$20,000. Related to the land issue and presenting another interesting paradox is how locals have found employment by looking after the homes of foreigners who have fled. They now live as caretakers in homes owned by foreigners, built by local labor on land sold by locals. This gives them a stable income and free place to stay. They are not complaining! This opportunity, however, is a result of increased crime along the coast. As the economic crisis deepens, house robberies and personal assaults have unfortunately increased.

Fishing has boomed. Fisherfolk that had turned away from the sea for tourism dollars have returned to the sea. They fish for export and to provide local markets with fresh fish. Other lifelong fishermen, like Manca Ruiz (Fig. 5.2) who have always relied on fishing for both local and export markets, continue to fish in the manner they have always done, but now also welcome locals into their team who have lost their jobs in the tourism industry (Fig. 5.3). The government has aided their activities by providing tax-free gasoline and discounted prices on outboard engines. In addition to providing economic stimulus to the fishing economy, local fishermen say the government is now even less likely to enforce regulations related to the use of the Marine Protected Area and have also looked the other way as illegal sea turtle egg harvesting has increased. Manca stated simply, "how can the government deny food to the people?" "Turtle eggs are food and people are hungry."

Gigante fishermen also state that ad hoc sales of fish to tourists (often lobster that did not meet size requirement for international markets) and Gigante restaurants have declined. Additionally, many of the fisherfolk who converted fishing businesses to other forms of tourism-oriented businesses during the boom have found it difficult to re-enter the fishing business after having sold their nets and other fishing equipment. Nets, line, and hooks are not cheap.

Many ex-employees of tourism businesses have now set up their own small businesses. Two long-term Gigante residents, Dona Isolina and Don Goyo have set up tiny ramshackle restaurants to serve the few beach visitors (yes, literally made up of driftwood and cooking on open wood fires). They, and other small operations like them, serve the surfing and backpacking visitors. Surfers and backpackers continue to visit Nicaragua, albeit in highly reduced numbers. Surfers revel in the empty waves and backpackers take joy in the reduced prices for all services. It seems then that

Fig. 5.2 Manca Ruiz has fished the seas around Gigante since the early 1990s. He has been somewhat immune to changes brought about by the decline in tourism because he has always relied on the sea rather than tourism. *Source* Jamey Thomas

those who "founded" southwest Nicaragua's tourism scene, the surfers, are those who continue to sustain the tourist economy at a base level. They just keep it ticking over.

5.3 Concluding Thoughts

All of the local residents interviewed for this chapter lamented the political crisis and the downturn of their personal economies. Understandably, they are afraid to speak out openly against the government. But, in careful words, they express the need for a rapid change in government that will bring tourists back. They fear though that the damage to its image might take time for Nicaragua to repair—they are very aware it took a long time for Nicaragua to overcome the legacy of war as tourism grew in the 1990 and 2000s. Many wonder if tourism will recover to the heights of 2017 and early 2018. In the meantime, locals remain resourceful. Many are now surviving together by pulling on their years of social capital formed during the revolution and formation of peasant cooperatives in the area.

Clearly, the decline in tourism has had drastic economic impacts on them, while simultaneously relieving pressure on conflicts related to fisheries, groundwater,

Fig. 5.3 Fishermen like Manca have incorporated new members into their fishing teams. *Source* Jamey Thomas

and land. The downturn in tourism undoubtedly results in diminished demands for groundwater from that sector and might lead to the conclusion that access to freshwater will improve for local residents. In reality, the picture is more complicated. As pointed out in Chap. 3 and in this chapter, the boom of tourism from 2010 to 2018 coincided with a period of below average rainfall. Thus, it was difficult to determine if the drop in the water table (and subsequent drying of wells) had more to do with over abstraction of groundwater to meet tourism demand or with the lack of rainfall to sufficiently recharge aquifers. Further, the new period of economic and political crisis that has brought the dramatic decline in tourism overlaps with a return to average rainfall and recharging of aquifers. Again, these trends make it challenging to disentangle culpability. The only way to determine specific causation between the tourism industry and local water security is to extend the period of research on consumption, rainfall, and recharge through the various cycles of tourism and rainfall patterns. Future research will surely need to capture this and other parts of the story of resources, local populations, and the extensive reach of tourism—for better or worse.

References

Alvarez W (2019) Nicaragua pierde 433.6 millones de dólares en inversión extranjera y por la caída del turismo entre enero y marzo de 2019. *La Prensa*, July 25, 2019. https://www.laprensa.com.ni/2019/07/25/economia/2573004-nicaragua-pierde-433-6-millones-de-dolares-en-inversion-extranjera-directa-entre-enero-y-marzo-de-2019. Accessed 26 August 2019

Martí S (2018) Nicaragua: continuities and ruptures of a crisis. The Global Americans, August 2, 2018. https://theglobalamericans.org/2018/08/nicaragua-continuities-and-ruptures-of-a-crisis/. Accessed 26 August 2019

The manufacturer's authorised representative in the EU is Springer
Nature Customer Service Centre GmbH, Europaplatz 3, 69115 Heidelberg,
Germany. If you have any concerns regarding our products, please
contact ProductSafety@springernature.com

Printed and bound by CPI Group (UK) Ltd, Croydon, CR0 4YY
30/09/2026
02225664-0005